Mesoscopic Physics meets Quantum Engineering

LECTURE NOTES

Mesoscopic Physics meets Quantum Engineering

Sergey N. Shevchenko

B. Verkin Institute for Low Temperature Physics and Engineering of the National Academy of Sciences of Ukraine & V. N. Karazin Kharkiv National University

World Scientific

NEW JERSEY · LONDON · SINGAPORE · BEIJING · SHANGHAI · HONG KONG · TAIPEI · CHENNAI · TOKYO

Published by

World Scientific Publishing Co. Pte. Ltd.

5 Toh Tuck Link, Singapore 596224

USA office: 27 Warren Street, Suite 401-402, Hackensack, NJ 07601

UK office: 57 Shelton Street, Covent Garden, London WC2H 9HE

British Library Cataloguing-in-Publication Data

A catalogue record for this book is available from the British Library.

MESOSCOPIC PHYSICS MEETS QUANTUM ENGINEERING

ISBN 978-981-120-139-4

For any available supplementary material, please visit
https://www.worldscientific.com/worldscibooks/10.1142/11310#t=suppl

Typeset by Stallion Press
Email: enquiries@stallionpress.com

Abstract

Quantum mechanics was initially constructed to describe objects on atomic and subatomic scales. However, in the last decades, quantum mechanics has been revisited and its use extended to the study and description of macroscopic distinct states. This is accomplished by modeling basic objects of mesoscopic physics, such as superconducting quantum circuits and low-dimensional structures derived from a two-dimensional electronic gas. In recent years, these devices support the study of fundamental systems such as a two-level quantum system, or qubit, as an object for manipulations and applications. This book will provide an introduction to quantum computation and quantum information, based on quantum physics, solid-state theory, and theory of computing. We will become familiar with this important field and explore how it is inseparably linked to basic notions of physics such as superposition, entanglement, and quantum dynamics. Then we will consider superconducting and mesoscopic systems, as well as a series of phenomena, where important are the spectra quantization, interference, and charge discreteness.

The contents of this book are based on several sources. The references are cited as footnotes in the text, while the most important references (recommended for further reading) for this course are given as a separate list at the end. An asterisk (*) indicates paragraphs which contain supplementary information and can be omitted for a first reading. Problems are listed at the end of each chapter, and these can be used for either homework or self-study. The difficulty of these problems is indexed by the number of asterisks, from one to five.

The lecture course from which this book derives its content is intended for graduate students and postdocs who are acquainted with quantum mechanics and statistical physics. In particular, it was developed while delivering lectures to the 5th year students of the Department of Physics and Technology in Kharkov National University.

The aims of this book are:

- To expand and deepen readers' understanding of quantum mechanics and solid-state physics. In particular, to teach the theoretical basics in fields such as quantum computation, quantum engineering, and circuit quantum electrodynamics;
- To familiarise readers with common topical problems in mesoscopic physics and mesoscopic objects, such as superconducting quantum circuits, low-dimensional conductors, electric and nanomechanical resonators;
- To examine simple models that describe realistic systems in quantum engineering.

Contents

Chapter 0

QUANTUM ENGINEERING

"Today there is a substantial white spot in the physical picture of the world, namely: a bridge is absent between the submicroscopic level of quantum mechanics and macroscopic world of classic physics."

[Penrose 2003]

0.1. Mesoscopics: why it is important

In this lecture course, we are mostly concerned with objects of size much larger than that of microscopic atomic systems but smaller than that of the usual macroscopic things around us. We are used to the fact that phenomena in microscopic atomic world are described by quantum mechanics while those on macroscopic scales are described by classical physics. But where exactly is the border between them and how do we describe the intermediate region? Numerous modern researches are devoted to such questions, forming the basis of so-called mesoscopic physics, or mesoscopics, for brevity. The formulation and solution of new problems in quantum mechanics, which is more than a century old, is a "quantum challenge" for contemporary researchers [Greenstein and Zajonc 2006].

Accordingly, we face the necessity of introducing a new lecture course into university programs for students in physics. To date, there is no established and standard lecture course in mesoscopic physics; the issue of which phenomena and effects should be included is still left to the lecturer's discretion. Therefore, usually the lecturers

1

of their respective courses focus on the fields in which they are experts, e.g.,[1] [Moskalets 2010] and [Zagoskin 2011]. Alternatively, other standard courses of physics may be modified to include the study of mesoscopic systems. The author of the present course, albeit also based on personal experience and understanding, has included the descriptions of phenomena, problems, and techniques which are useful for the education and further research work of students in physics, and which are topical to date. The following subsection describes several examples of mesoscopic phenomena.

Now, let us try to build an impression about mesoscopics. This word originates from the Greek "$\mu\varepsilon\sigma o\zeta$" — intermediate.[2] This is the field of physics which describes phenomena on intermediate scales between the microscopic and macroscopic. Namely, *mesoscopic physics* studies the appearance of quantum effects in many-particle systems (in condensed media) when the system phase coherence is relevant. Here quantum coherence assumes continuous wave-function phase changes, which result in quantum interference phenomena. Losing the phase results in decoherence. Until there is phase coherence, the system containing any number of particles is quantum in behavior. Characteristic scales are the *decoherence length* L_φ and *the decoherence time* T_φ. The success of countless experiments in increasing these values has led to the possibility for observing and discussing mesoscopic phenomena. Of course, the values of these characteristic parameters depend on a physical system, but in general, we can call a system mesoscopic if its size (at least in one dimension) is much larger than the atomic size and much smaller than the decoherence length L_φ, and a characteristic evolution time is much smaller than the decoherence time T_φ. As an initial approximation, L_φ and T_φ values are on the scale of a micrometer and microsecond, respectively.

In relation to the topicality of studying mesoscopic systems, in addition to the above citation from the Penrose book, we quote

---

[1]Y. Imry, Introduction to Mesoscopic Physics, Oxford University Press (2001).

[2]N. G. van Kampen, The expansion of the master equation, Adv. Chem. Phys. **34**, 245 (1976).

[Nielsen and Chuang 2010]: "What is it that separates the quantum and the classical world? What resources, unavailable in a classical world, are being utilized in a quantum computation? Existing answers to these questions are foggy and incomplete; it is our hope that the fog may yet lift in the years to come, and we will obtain a clear appreciation for the possibilities and limitations of quantum information processing." And even though much time has passed since the formation of quantum mechanics, this thesis has something in common with Niels Bohr's statement: "Anyone who is not shocked by quantum theory has not understood it."

0.2. About the structure of this lecture course

(i) In his most cited paper,[3] the most famous physicist, A. Einstein, together with B. Podolsky and N. Rosen, discusses the quantum correlations and the entanglement. Another famous physicist, R. Feynman, in his most cited article[4] proposes to use these correlations for simulating quantum processes and for building a quantum computer. The experimental possibilities in the last two decades allow such speculative discussions to be translated to detailed and concrete scientific research. Quantum mechanics becomes a working tool of physicists when dealing with mesoscopic-size systems... Usually, standard university courses in physics do not mention this, when in fact even what we have discussed should be enough stimulus for introducing new separate lecture courses. In our case, such speculations justify Chapter 1.

(ii) Motivations of research in mesoscopic physics are multifaceted: they arise from the necessity of developing the elementary basics for microelectronics as well as by the desire to find answers for gnosiological questions. The study of mesoscopic systems allows quantum mechanics to be understood more

---

[3]A. Einstein, B. Podolsky, and N. Rosen, Can quantum-mechanical description of physical reality be considered complete?, Phys. Rev. **47**, 777 (1935).

[4]R. P. Feynman, Simulating physics with computers, Int. J. Theor. Phys. **21**, 467 (1982).

deeply and broadly. With that, what is most important here is the dynamic behaviour of such systems, as stated in the conclusion of the review article [Valiev 2005]: In addition to well-studied statics of quantum systems, mesoscopic physics adds new aspects of dynamics to the field of quantum mechanics, such as the regime of strong driving and weak measurements. Besides, while it is important to have fine control and measurements over mesoscopic systems, these are inevitably connected to the dissipative environment. We will discuss the dynamic behaviour of driven dissipative quantum systems in Chapter 2.

(iii) We should note that the coherent effects in quantum systems have already been known for quite a long time. One of the most interesting phenomena, where quantum laws appear at the macroscopic scale, is superconductivity. Moreover, in effects such as flux quantization and the Josephson effect, the wave-function phase of the superconducting condensate is important. Therefore, such effects are called "macroscopic" quantum or coherent effects. We have to separate these effects from "really quantum" effects, where instead the superposition of macroscopically distinct states is important. We will consider these issues in Chapter 3, beginning with basic classical superconducting systems and ending with their quantum counterparts, the superconducting qubits.

(iv) One of the most striking quantum effects is the Aharonov–Bohm effect, where the changes of a wave function are defined by the magnetic-field vector-potential and not by the magnetic field itself. For multiply-connected mesoscopic normal samples, this results in the oscillatory dependence of the conductance on the magnetic flux. In particular, this results in the appearance of the so-called persistent current in a normal-metal ring pierced by magnetic flux. Such interference can bear both constructive and destructive character. This means that the current through the circuit of several mesoscopic conductors cannot be described by the aggregation of successive and parallel resistances. Such a classical interpretation does not take inteference into account. Also, the conductance itself does not follow Ohm's law, but rather is described in terms of the conductance quantum. Thus, the engineering of mesoscopic normal circuits differs in principle from electronics of classical circuits. We will discuss this in Chapter 4.

(v) Mesoscopic systems are interesting because they have parameters which are tunable over a wide range, and also because they can be integrated with other systems. These could be diverse systems: researchers study the connection of mesoscopic systems with single atoms, with other mesoscopic systems, or with macroscopic circuits. Then they form compound systems. They are also called hybrid systems, if the subsystems are diverse. Here an important example is the connection with a measuring circuit. In Chapter 5 we will consider coupling with characteristic systems such as electric and nanomechanical resonators. The manner by which such systems can be described — in terms of the dressed states when the resonator is quantum, or in the framework of the semiclassical approximation when it is classical — will be shown.

0.3. Fundamental and applied aspects of mesoscopics

As stated above, to a large degree, the emergence and the development of mesoscopics may be attributed to progress in technology and measurement techniques at the nano- and micro-scales. On the other hand, these mesoscopic systems are an interesting avenue for the creation of new electronic devices. That is why the field of study of mesoscopic systems is also called *quantum engineering.*

It is interesting that new technological achievements of the previous century were built also on the laws of quantum physics, but are exploited as classical devices: lasers and transistors are based on knowledge of the quantum spectra of gases and solids; atomic energetics is based on quantum atomic physics. At that, the phenomena and values which are observed and utilized (the current of electrons and the flow of photons) include large numbers of quantum particles, of which the averaged behaviour is described by classical currents, voltages, and electromagnetic waves. The progress of the last decades allows for the discussion of the concept of quantum devices, which are not only based on the laws of quantum physics, but also function in the quantum regime.

At present, a great many quantum technologies are being developed, and we will cover them in the following chapters. To date, some of them have resulted in the creation of commercial devices, for example, in quantum cryptography. Another important example (to date, arguably hypothetical) is the quantum computer. Will it be realized and for what will it be used? This is an open question (which will also be touched on in the next Chapter). But for us, and researchers in general, it is more important to note that such ideas have stimulated and continue to stimulate deeper and broader study of fundamental physical phenomena.

Of less interest to the public, but more interesting for physicists are the devices, which use new research findings and create the basis for further exploration. In this sense, mesoscopic devices open up unique possibilities: they work according to quantum laws, but have variable parameters and can be connected to macroscopic systems. One of such illustrative tools, which we will consider in detail, is Landau–Zener–Stückelberg–Majorana interferometry. It will be shown how the effects, which at the advent of quantum mechanics were considered by the classicists for microscopic quantum systems, can be realized with modern means and used to widen the quantum toolbox.

Chapter 1

SUPERPOSITION, ENTANGLEMENT, AND QUANTUM COMPUTATION

> "One of the goals of quantum computation and quantum information is to develop tools which sharpen our intuition about quantum mechanics, and make its predictions more transparent to human minds."
>
> [Nielsen and Chuang 2010]

The last three decades bore witness to the emergence of the new research field, Quantum Computations (also known as Quantum Information). *Quantum Computations* assumes the study of problems related to the processing and transfer of information using the laws and objects of quantum mechanics. The theory of quantum information appeared at the intersection of earlier fields such as information theory, computer sciences, and quantum mechanics. The formalism and methodology of quantum optics, condensed matter theory, and cryptography are also incorporated.

As we will discuss below in more detail, the long-standing goal of the theory of quantum computations is the development of a quantum computer.[5] The realizations of the quantum search algorithm, quantum cryptography, and quantum simulators are intermediate-scale problems researchers are considering.[6]

---

[5]T. D. Ladd, F. Jelezko, R. Laflamme, Y. Nakamura, C. Monroe, and J. L. O'Brien, Quantum computers, Nature **464**, 45 (2010).

[6]I. Buluta and F. Nori, Quantum simulators, Science **326**, 108 (2009); I. Georgescu, S. Ashhab, and F. Nori, Quantum Simulation, Rev. Mod. Phys. **86**, 153 (2014).

On the other hand, tasks and problems formulated by quantum computation theory stimulate the development and broadening understanding of quantum physics laws. It is this last thesis of quantum information theory that is the locomotive of the present lecture course. In this chapter we will give an introduction to quantum information theory, and, using the language of this theory, show how the basic notions of quantum mechanics can be described.

1.1. Quantum computers

Even though different aspects of quantum information theory are present in quantum physics from its very foundation, it is common to consider the R. Feynman talk at the conference some 35 years ago to be the beginning (see Footnote 4 on page 3; note that the idea of quantum computations was proposed by Yu. Manin in 1980[7]). In that speech, Feynman discussed principal difficulties with simulating quantum-mechanical systems with a usual (classical) computer. For that, he proposed to construct principally new computers, based on the laws of quantum mechanics. This thesis was later strictly grounded and developed. Maybe, it will be this very application — the simulation of quantum systems — that will become an important realization of quantum computations (see Footnote 6 on page 7).

Another reason leading to the search for new principles of calculations, is the shift in size of manufacturing elements of classical computers to nanometer scales, where the laws of quantum mechanics are relevant. Impressively, the development of computer technology during the last fifty years obeyed Moore's law with high accuracy. This law predicted the doubling of computation power for the same price every two years. To date, the progress was mainly due to miniaturization of the elements. And now, when the characteristic scales have shrunk down to the order of a few nanometers, quantum phenomena must inevitably be taken into account. However, quantum interference and fluctuations are ruinous for the operation principles of a classical computer. Thus the necessity in searching

---

[7]Yu. I. Manin, *Computable and Noncomputable*, Moscow, Soviet Radio, 1980.

for new models of informatization arises, and so the paradigm of a quantum computer appeared. Despite the existing series of problems of principle, many specialists believe that a quantum computer (if ever created) will be more powerful than any imaginable computer running on classical principles (see Footnote 5 on page 7).

The supremacy of a quantum computer, to date, is rather a technological issue. But there is another cause for studying the supremacy of quantum computations. There are problems which a quantum computer can solve effectively and a classical computer cannot. In this context, the term *effective* calculations assumes the algorithms which are realized for the times that grow polynomially with an increase in the system; non-effective algorithms are realized on the exponentially growing times. To date, three classes of problems have been developed where the quantum computer is most effective. First is the aforementioned *quantum simulation*. (For more information on the recent progress in this direction of applications to quantum chemistry, see Footnote 8.)

The second class of problems for quantum computers is the *decomposition of integer numbers* in prime factors. The foundation for this was proposed by D. Deutsch, and the corresponding algorithms were developed by P. Shor. The third circle of problems for a quantum computer was described by L. Grover — these relate to the *search* in non-structured environment [Valiev 2005].

In the present course we will deal only shortly with algorithms for quantum computers and we will not address many problems of quantum information, such as error corrections. The goal of our introduction is to consider conceptual points of quantum computation theory.

So, what is a quantum computer? In the simplest approach, a quantum computer contains a register of n qubits, controlled by means of external classical pulses [Valiev 2005], see Fig. 1.1. The controlled evolution of qubits state corresponds to the execution of algorithms. This evolution from a state ψ_i to a state ψ_f is described mathematically by a unitary matrix U of dimensionality $2^n \times 2^n$.

---

[8]A. Kandala *et al.*, *Nature* **549**, 242 (2017).

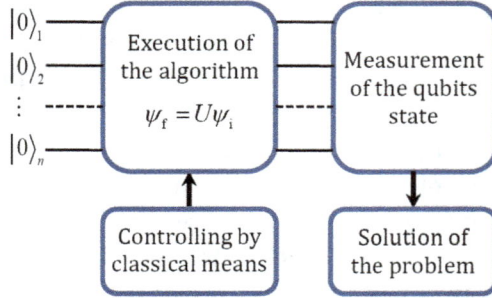

Fig. 1.1.   Schematic of a quantum computer.

After execution of an algorithm, the solution of the original problem is defined from the measured state. In the following we will consider this in more detail, but let us first discuss the notion of a qubit.

1.2. Qubits and superposition

A quantum bit, or a qubit, is in principle any quantum system with two possible states. The general theory of quantum computation and information is built on the notion of an abstract qubit, without detailing its physical origin.

Analogously to how a bit is described by states 0 and 1, a qubit is described by vector-states $|0\rangle$ and $|1\rangle$. The qualitative distinction is the ability of a qubit to be in a *superposition state*:

$$|\psi\rangle = \alpha |0\rangle + \beta |1\rangle. \tag{1.1}$$

This can be interpreted as the ability of a qubit to be in states 0 and 1 simultaneously. Two qubits can simultaneously take four values — 00, 01, 10 and 11. Each additional qubit doubles the number of possible states. For n qubits there are 2^n possible states. And a quantum register of only 350 qubits can support 2^{350} values simultaneously. This is more than the number of atoms in the visible part of the universe, and more than the so-called googol, quantifying 10^{100}.

The basis states of a qubit can be written in the form

$$|0\rangle = \begin{pmatrix} 1 \\ 0 \end{pmatrix}, \quad |1\rangle = \begin{pmatrix} 0 \\ 1 \end{pmatrix}, \tag{1.2}$$

and then $|\psi\rangle = (\alpha, \beta)^T$. As a graphic illustration, we can use the so-called *Bloch sphere*, expressing the coefficients α and β through the angles:

$$\alpha = \cos\frac{\theta}{2}, \quad \beta = e^{i\phi}\sin\frac{\theta}{2}. \tag{1.3}$$

(The common phase factor is not written here since it does not result in any observable consequence.) We can see that the poles, $\theta = 0, \pi$, correspond to basis states and the equatorial plane is described with $\theta = \pi/2$, which corresponds to the states equidistant from the poles, where $|\alpha| = |\beta|$. Then the vector-state evolution is described by showing the trajectory on a sphere of unit radius in terms of the polar and azimuthal angles, Fig. 1.2. A given state $|\psi\rangle$ can be described as a consequence of two rotations from an initial state, say, from a "north pole":

$$|\psi\rangle = \begin{pmatrix} 1 & 0 \\ 0 & e^{i\phi} \end{pmatrix} \begin{pmatrix} \cos\frac{\theta}{2} & -\sin\frac{\theta}{2} \\ \sin\frac{\theta}{2} & \cos\frac{\theta}{2} \end{pmatrix} |0\rangle. \tag{1.4}$$

The complex coefficients α and β can, in principle, take any two values. From such a wealth of possibilities follows practical interest in

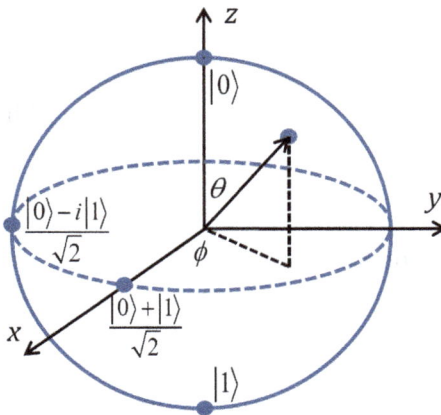

Fig. 1.2.   Bloch sphere, demonstrating values of the vector-state.

qubits. But can we make qubits macroscopic? This problem will be discussed later. Now, in relation to the essence of superposition and in relation to using quantum mechanics for description of classical systems, let us consider the famous Schrödinger's cat paradox. This describes a cat in a box where there is a poison linked to a radioactive atom. The cat is assumed to be alive until the atom spontaneously decays, triggering the poison which kills the cat. In such a formulation, according to the laws of quantum mechanics, a system (the cat plus the atom) isolated from the external world is in the quantum superposition of the two states:

$$|\Psi\rangle = \alpha |0, \uparrow\rangle + \beta |1, \downarrow\rangle. \qquad (1.5)$$

Here the states $|\downarrow/\uparrow\rangle$ correspond to the "ground" (dead) or "excited" (alive) state of the cat at the decomposed atom ($|1\rangle$) and not-decomposed atom ($|0\rangle$), correspondingly. Opening the box and executing the measurement, we observe either a live or dead cat. The vector-state is reduced to one of the components of the superposition state. The paradox is whether the cat is alive or dead *before* the measurement of the system state. From the point of view of quantum mechanics (in its Copenhagen interpretation) there is no paradox. The question of which state the system is in before the measurement is forbidden. In the example of Schrödinger, formulated so dramatically, the uncertainty was initially limited by the atom size and then was extended to the macroscopic level, and this uncertainty is resolved by direct observation. Even Schrödinger recognizes that such a model of reality does not contain anything ambiguous or controversial.

The fundamental problem is whether we can extend the principle of quantum superposition to the macroscopically distinct states. Applying classical description to a quantum system and ignoring fundamental non-classical states result in paradoxical conclusions. At present, it is clear that these paradoxical conclusions characterize an unusual quantum reality that, nevertheless, takes place. And indeed, recently the macroscopically distinct superposition states, also known as the Schrödinger-cat states, were observed in diverse systems. We will also discuss these later.

Let us now discuss some possible physical realizations of qubits on the basis of microscopic quantum systems.

1.3. Physical realizations of qubits

1.3.1. *Requirements for the candidates in qubits*

A physical system which is considered as a candidate for the role of a qubit has to fulfill some criteria, compactly formulated as the *DiVincenzo criteria*:

(1) Scalability. We need to have a scalable system of qubits with well-defined parameters and with the ability to create the entangled states. In particular, this means that the upper levels of the physical realization of the qubit device are well separated from the operational two levels.

(2) Initialization. There should be the ability to prepare given states. For computations, this initial state can be the ground state. Then, this can be reached by the cooling in practice.

(3) Isolation. Good isolation from the environment and large decoherence times are needed. These times should be at least three orders of magnitude larger than a characteristic time needed for an operation, to have the ability to work with the information before it is lost, to transmit the information and to initialize the quantum error correction algorithms.

(4) Control. We should have the ability to make one- and two-qubit unitary operations. It was proven that these are sufficient for all the problems on multi-qubit systems.

(5) Measurement. For finalizing quantum algorithms, we need the capability to reliably measure the states of individual qubits.

In practice, however, we have to strike a balance among full realizations of each criterion, since, for example, the last criterion requires the read-out electronics, while the third criterion assumes maximal isolation from the environment.

To date, there is a large number of different systems proposed for the role of qubits. These systems can be microscopic two-level systems, such as electronic states or photon polarizations, as well as artificial mesoscopic systems such as quantum dots and super-conducting circuits. All these systems possess their own advantages and disadvantages from the point of view of realizing controllable two-level systems.

Microscopic systems have long coherence times, but it is difficult to control them individually, to write and read out information. Parameters of such systems, as a rule, are defined during the manufacture of these systems and cannot be changed in the working process of the device. Mesoscopic artificial circuits can be prepared with the predetermined parameters, while their working parameters can be changed during their deployment by means of external control-ling currents or voltages. We will consider this in more detail using the example of superconducting qubits. In contrast to microscopic qubits, there are difficulties with isolating their mesoscopic analogues from the environment.

1.3.2. *About the control and manipulation of individual quantum systems*

Here it is appropriate to become familiar with two formulations of problems related to modern prospects of working with micro-scopic systems. Namely, consider the works of two groups of the 2012 Nobel Laureates — Serge Haroche and David Wineland. The former studies the states of photons by means of atoms, while the latter studies the states of atoms by means of photons. Such formulation is characteristic of quantum optics, which is the field studying the interaction of atoms and fields, matter and light. The Nobel prize was awarded "for ground-breaking experimental methods that enable measuring and manipulation of individual quantum systems". Here we can clarify that this refers to microscopic quantum systems. And in the present course, we will largely con-sider how to probe and manipulate individual *mesoscopic* quantum systems.

Of interest in our current discussion is an opinion expressed by the Nobel committee regarding the prospects of this research. Quoted verbatim: "Their methods have enabled this field of research to take the very first steps towards building a new type of super-fast computer based on quantum physics. Perhaps the quantum computer will change our everyday lives in this century in the same radical way as the classical computer did in the last century."

At present, researchers study not only individual qubits of different kinds, but also hybrid systems, which include subsystems of diverse types.[9] In such composite systems, diverse elements can be used for writing, saving, treating and reading-out the information.

Among the realizations of two-level systems we can single out such fundamental objects as spins, photons and two-level atoms. We shall now briefly consider such microscopic realizations of qubits as a repetition of what is known, while the mesoscopic systems will be considered later in detail, in the following chapters.

1. *Spin and photon*

Let us give a refresher on particle spin in magnetic field, known from the course of quantum mechanics (see Chapters 8 and 15 in [Landau and Lifshitz 1977]).

The Hamiltonian of a particle with a spin in the electromagnetic field has the form

$$\widehat{H} = \frac{1}{2m}\left(\widehat{\vec{p}} - \frac{e}{c}\vec{A}\right)^2 + e\varphi - \widehat{\vec{\mu}}\vec{H}, \tag{1.6}$$

where φ and $\vec{A}$ are the scalar and vector potentials of the field, $\vec{H}$ stands for the magnetic field, $\widehat{\vec{\mu}}$ is the magnetic-moment operator, corresponding to the spin, $\widehat{\vec{\mu}} = \frac{\mu}{s}\widehat{\vec{s}}$, with s and $\widehat{\vec{s}}$ being the value and the operator of the spin. In particular, for an electron, the ratio μ/s equals $-|e|\hbar/mc$, which means that the electron intrinsic magnetic moment equals, up to the sign, to the Bohr magneton, $\mu_B = \hbar|e|/2mc$. This can be rewritten, introducing the Landé g-factor, which for an electron equals approximately to 2: $\mu = -g\mu_B s = -\mu_B$.

Usually we can factorize a particle wave function into the coordinate and spin parts. Being not interested in the free-motion

---

[9]Z.-L. Xiang, S. Ashhab, J.-Q. You, and F. Nori, Hybrid quantum circuits: superconducting circuits interacting with other quantum systems, Rev. Mod. Phys. **85**, 623 (2013).

wave function, we consider only the spin part of the Hamiltonian. For a particle with spin $s = 1/2$, the spin operator is given by the Pauli matrices, $\widehat{\vec{s}} = \frac{1}{2}\widehat{\vec{\sigma}}$, and the Hamiltonian equals (omitting the hats, here and below):

$$H = \frac{1}{2}g\mu_B \vec{\sigma}\vec{H}, \tag{1.7}$$

$$\vec{\sigma} = (\sigma_x, \sigma_y, \sigma_z), \quad \sigma_x = \begin{pmatrix} 0 & 1 \\ 1 & 0 \end{pmatrix},$$

$$\sigma_y = \begin{pmatrix} 0 & -i \\ i & 0 \end{pmatrix}, \quad \sigma_z = \begin{pmatrix} 1 & 0 \\ 0 & -1 \end{pmatrix}. \tag{1.8}$$

In the general case, it is interesting to formulate the problem, when there are constant and alternating components of the magnetic field. Choosing the former along the x axis and the latter along the z axis and introducing an evident change in notations, we obtain

$$H = -\frac{\Delta}{2}\sigma_x - \frac{\varepsilon(t)}{2}\sigma_z. \tag{1.9}$$

So, we have obtained the Hamiltonian describing spin $1/2$ particle. The spin wave function — the spinor — is the two-component column vector. As the basis, we can take, for instance, the eigenvectors of the σ_z operator. We can also find eigen-values of the Hamiltonian; there are two of them and they describe our two-level system, the qubit. We will consider this in more detail in the next Chapter.

An analogous Hamiltonian can be used to describe other two-level systems. Then the Hamiltonian is called pseudo-spin one. One example involves the energy levels of a two-atom molecule, e.g. the ammonia molecule. This is nicely described in §79 of [Landau and Lifshitz 1977] and in Chapter 7 of the Feynman lectures.[10]

Consider now another (flying) qubit — a photon. As we will see, this also can be considered as a particle with pseudo-spin $1/2$, see

---

[10]R. Feynman, R. B. Leighton, and M. L. Sands, The Feynman Lectures on Physics, Vol. III, Quantum Mechanics, Addison-Wesley, MA (1965).

for more detail §1.2 in [Blum 1981] and in the review article by Klyshko.[11]

Let us start here from the electromagnetic field of a plane monochromatic wave travelling along the z axis:

$$\mathbf{E} = \mathrm{Re}\,A\mathbf{e}\exp\left(ikz - i\omega t\right), \tag{1.10}$$

where A is the wave amplitude and k is the wave number, which is related to the wavelength Λ, $k = 2\pi/\Lambda$. Polarization unit vector (Jones vector) $\mathbf{e}$ is perpendicular to the direction of the wave propagation, so it can be decomposed into two components:

$$\mathbf{e} = \alpha\mathbf{e}_x + \beta\mathbf{e}_y,$$
$$\alpha = \cos\frac{\theta}{2}, \quad \beta = e^{i\varphi}\sin\frac{\theta}{2}. \tag{1.11}$$

We have presented the wave as a superposition of two plane-polarized waves, described by the unit vectors $\mathbf{e}_x$ and $\mathbf{e}_y$. The coefficients are chosen so that the absolute value of the vector $\mathbf{e}$ is unity, $\mathbf{e}\mathbf{e}^* = 1$. In particular, at $\alpha = \beta = 1/\sqrt{2}$, the photon is polarized at angle $\pi/4$ to the x axis, and with $\alpha = 1/\sqrt{2}$ and $\beta = \pm i/\sqrt{2}$ it has right and left polarizations, respectively. The latter case corresponds to $\varphi = \pm\pi/2$ and $\theta = \pi/2$, and it gives $\mathbf{e} = \left(\mathbf{e}_x \pm i\mathbf{e}_y\right)/\sqrt{2}$. In the general case we have elliptic polarization.

For a single photon, one can define the state vector $|e\rangle$, which can be decomposed in the basis vectors, corresponding to the polarizations along the x and y axes:

$$|e\rangle = \cos\frac{\theta}{2}\,|e_x\rangle + e^{i\varphi}\sin\frac{\theta}{2}\,|e_y\rangle. \tag{1.12}$$

This vector is fully analogous to a qubit, which is a system with the vector-state defined in Eq. (1.1). One can visualize a state of a photon-qubit by choosing the parameters φ and θ as the azimuthal and polar angles, respectively. Then the states with different polarizations will be described by points on the unit sphere

---

[11]D. N. Klyshko, Basic quantum mechanical concepts from the operational viewpoint, Phys. Usp. **41**, 885 (1998).

which is called Poincaré sphere. This is analogous to the Bloch sphere considered above.

Here we have to clarify that, strictly speaking, a photon cannot be ascribed a notion of spin as the full angular momentum at rest, since its mass at rest equals to zero. Formally, however, a spin equal to unity can be ascribed to a photon, since its wave function is described by the vector (1.10). But since the magnetic field is perpendicular to the propagation direction, which is along the z axis, the projection of spin in perpendicular direction is excluded, and there are only two possible values for the spin direction, along and against the propagation. These are characterized by the value $\lambda = \pm 1$, which is called the *spirality* of a photon. The spirality equals the projection of a full moment $\mathbf{J} = \mathbf{L} + \mathbf{S}$, consisting of the orbital moment and the spin, to the z axis: $\mathbf{J}\mathbf{e}_z = (\mathbf{L} + \mathbf{S})\,\mathbf{e}_z = \mathbf{S}\mathbf{e}_z = \lambda$. From quantum electrodynamics we know that the states with $\lambda = \pm 1$ correspond to the light with circular polarization, which is called right/left polarization respectively, and they are described by the vector-state $|\pm 1\rangle$. As we have seen, $|\pm 1\rangle = \mp (|e_x\rangle \pm i\,|e_y\rangle)/\sqrt{2}$.

So, formally one can match the states of definite spirality $|\pm 1\rangle$ with the spin $1/2$ particle. These vectors can be described by the two-component columns, see Eq. (1.2). This basis can be used to decompose an arbitrary state: $|e\rangle = \alpha\,|+1\rangle + \beta\,|-1\rangle$. Thus, a photon can be a (flying) qubit, of which the role of the quasi-spin is fulfilled by the polarization.

1.4. Quantum logic operations

1.4.1. *One-qubit operations*

The coefficients α and β in (1.1) satisfy the condition of the normalization $|\alpha|^2 + |\beta|^2 = 1$. This condition also has to be satisfied by $|\psi'\rangle = U\,|\psi\rangle = \alpha'\,|0\rangle + \beta'\,|1\rangle$. From here it follows that $\langle\psi'\,|\psi'\rangle = \langle\psi|\,U^\dagger U\,|\psi\rangle = 1$. That is why the evolution is described by a unitary operator: $U^\dagger U = I$, where $U^\dagger = (U^T)^*$. The unitarity is the only restriction for the quantum operations. This means that any unitary matrix defines some quantum operation. This is in contrast with the classical case, where there is only one nontrivial one-bit operation, NOT ($0 \to 1$ and $1 \to 0$).

To the most important quantum operations we can attribute the following: X (or NOT), Y, Z — they coincide with the respective Pauli matrices —

$$X \equiv \sigma_x = \begin{pmatrix} 0 & 1 \\ 1 & 0 \end{pmatrix}, \quad Y \equiv \sigma_y = \begin{pmatrix} 0 & -i \\ i & 0 \end{pmatrix},$$

$$Z \equiv \sigma_z = \begin{pmatrix} 1 & 0 \\ 0 & -1 \end{pmatrix}, \tag{1.13}$$

and also the Hadamard operation (H), phase shift (S) and the so-called $\pi/8$ element (T):

$$H = \frac{1}{\sqrt{2}} \begin{pmatrix} 1 & 1 \\ 1 & -1 \end{pmatrix}, \quad S = \begin{pmatrix} 1 & 0 \\ 0 & i \end{pmatrix}, \quad T = \begin{pmatrix} 1 & 0 \\ 0 & e^{i\pi/4} \end{pmatrix}. \tag{1.14}$$

One can see that $H = \frac{X+Z}{\sqrt{2}}$, $S = T^2$, and the name of the T matrix becomes clear, if we take the factor $e^{i\pi/8}$ out of it. The Hadamard operation is also called "square root from NOT", since it transforms $|0\rangle \to (|0\rangle + |1\rangle)/\sqrt{2}$ — half-way between $|0\rangle$ and $|1\rangle$ (albeit $H^2 = I \neq X$). Note that the i-th basis vector has 1 in the i-th position. That is why the effect of the matrices on the basis vector $|0\rangle$ is defined by the first column, on the vector $|1\rangle$ by the second column, etc. The result of making the logic operations is illustrated in Fig. 1.3.

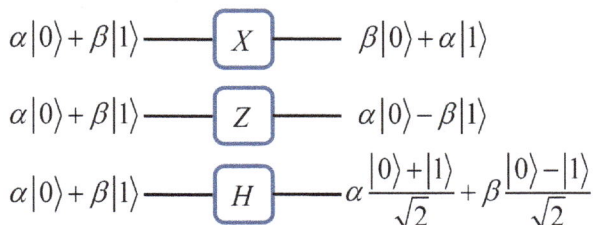

Fig. 1.3. Logic NOT, Z, and Hadamard operations.

1.4.2. *Controlled-NOT operation*

Any one-qubit operation can be realized using a finite set of quantum operations. More generally, arbitrary quantum computation on any number of qubits can be realized, using a final set of operations, which is called the *universal set* of operations for quantum calculations. There is the statement (In [Nielsen and Chuang 2010] §4.5) that *arbitrary operation on the space of states of n qubits can be realized using only one-qubit elements together with the controlled-NOT (CNOT) element.* We will consider the latter below. One of the possible universal sets — the so-called standard set — consists of the Hadamard, phase shift, $\pi/8$, and CNOT elements.

*In the theory of classical calculations there are 5 basic multi-bit classical operations: AND, OR, XOR (exclusive-OR), NAND (AND and NOT), and NOR (OR and NOT). These operations act on two bits and give one bit at the output. There is the statement: any function on bits can be calculated from the combination of only NAND operation, which is then called the universal gate (operation).

So, even though there are many interesting gates (operations), any multi-qubit operation can be composed from one-qubit operations and the CNOT operation. This latter two-qubit operation is defined by the impact of the first (control) qubit on the second (target) qubit, so that the value of the latter is changed only if the value of the former is 1. To be more specific, this reads:

$$|00\rangle \rightarrow |00\rangle \,, \quad |01\rangle \rightarrow |01\rangle \,, \quad |10\rangle \rightarrow |11\rangle \,, \quad |11\rangle \rightarrow |10\rangle \,. \qquad (1.15)$$

Or, shortly: $|A, B\rangle \rightarrow |A, B \oplus A\rangle$, where the symbol $\oplus$ denotes summation modulo 2; see Fig. 1.4.

It is useful also to write down the matrix presentation of this operation, U_{CNOT}. For this, let us first define the basis. Note that

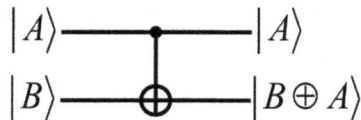

Fig. 1.4.　Controlled-NOT (CNOT) operation.

the vector-state for the two-qubit system is constructed by rules of the tensor product:

$$|\psi_1, \psi_2\rangle \equiv |\psi_1\rangle |\psi_2\rangle \equiv |\psi_1\rangle \otimes |\psi_2\rangle = \begin{pmatrix} \alpha \\ \beta \end{pmatrix} \otimes \begin{pmatrix} \gamma \\ \delta \end{pmatrix}$$

$$= \begin{pmatrix} \alpha \begin{pmatrix} \gamma \\ \delta \end{pmatrix} \\ \beta \begin{pmatrix} \gamma \\ \delta \end{pmatrix} \end{pmatrix} \equiv \begin{pmatrix} \alpha\gamma \\ \alpha\delta \\ \beta\gamma \\ \beta\delta \end{pmatrix}. \tag{1.16}$$

This means that the basis vectors have the form

$$|00\rangle \equiv |0\rangle |0\rangle \equiv |0\rangle \otimes |0\rangle = \begin{pmatrix} 1 \\ 0 \\ 0 \\ 0 \end{pmatrix},$$

$$|01\rangle = \begin{pmatrix} 0 \\ 1 \\ 0 \\ 0 \end{pmatrix}, \quad |10\rangle = \begin{pmatrix} 0 \\ 0 \\ 1 \\ 0 \end{pmatrix}, \quad |11\rangle = \begin{pmatrix} 0 \\ 0 \\ 0 \\ 1 \end{pmatrix}. \tag{1.17}$$

Again, the i-th basis vector has 1 in the i-th position. That is why the effect of U_{CNOT} on the first basis vector is defined by the first column, etc. We have

$$U_{\text{CNOT}} = \begin{pmatrix} 1 & 0 & 0 & 0 \\ 0 & 1 & 0 & 0 \\ 0 & 0 & 0 & 1 \\ 0 & 0 & 1 & 0 \end{pmatrix}. \tag{1.18}$$

*We can remark that the CNOT operation is the generalization of the XOR operation. But the other classical operations do not have any quantum analogue since they are the irreversible operations. Then, say, having $A \oplus B$ at the input, we cannot restore A and B from the output.

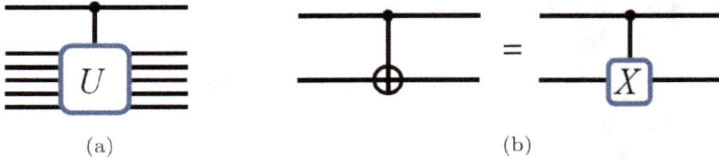

Fig. 1.5. Controlled U-operation (a) and its particular case, CNOT operation (b).

*Generalization of the CNOT operation on the three-qubit system with two control qubits and one target qubit realizes the Toffoli gate. This operation allows irreversible logic operations to be made and in this way classical calculations are realized on a quantum computer. So, a quantum computer can simulate (work as) a classical computer.

We can also introduce the generalization for the CNOT operation for the case of many qubits. In Fig. 1.5 the U-operation is demonstrated, with the particular case being the CNOT operation.

1.5. Quantum schemes

1.5.1. *Measurement and swap*

We consider the simplest quantum schemes, both to familiarize ourselves with the quantum calculations, and to understand the important link of this field to the foundations of quantum mechanics.

In Fig. 1.6 we present some of the basic quantum schemes, the ones for the measurement and for the swap operation. The measurement operation gives a classical bit M. For the state $|\psi\rangle = \alpha |0\rangle + \beta |1\rangle$ this bit is 0 with the probability $|\alpha|^2$ and 1 with the probability $|\beta|^2$. The scheme of the swap operation consists of triple CNOT operation

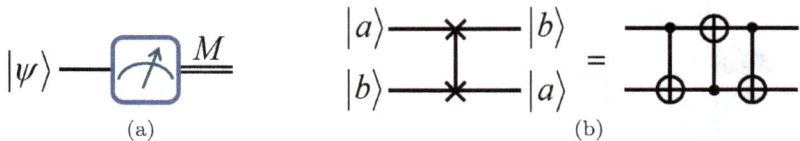

Fig. 1.6. (a) Operation of measurement and (b) the swap operation and its symbol.

and is described by exchanging the qubit's states, $|a, b\rangle \rightarrow |b, a\rangle$:

$$|a, b\rangle \rightarrow |a, a \oplus b\rangle \rightarrow |a \oplus (a \oplus b), a \oplus b\rangle$$
$$= |b, a \oplus b\rangle \rightarrow |b, (a \oplus b) \oplus b\rangle = |b, a\rangle. \qquad (1.19)$$

1.5.2. *No-cloning theorem*

Classical copying of a bit can be realized by means of the CNOT operations, as is shown in Fig. 1.7(a).

Let us now try to analogously copy the state $|\psi\rangle = \alpha |0\rangle + \beta |1\rangle$. Then we get

$$[\alpha |0\rangle + \beta |1\rangle] |0\rangle = \alpha |00\rangle + \beta |10\rangle \rightarrow \alpha |00\rangle + \beta |11\rangle. \qquad (1.20)$$

If we have $|\psi\rangle = |0\rangle$ or $|\psi\rangle = |1\rangle$, then this would give copying, but, in the general case, we expect at the output

$$|\psi\rangle |\psi\rangle = \alpha^2 |00\rangle + \alpha\beta |01\rangle + \alpha\beta |10\rangle + \beta^2 |11\rangle. \qquad (1.21)$$

It turns out that it is impossible to copy an arbitrary quantum state. (More precisely, the cloning operation can copy only orthogonal states, here – $|0\rangle$ and $|1\rangle$.) This statement is known as the no-cloning theorem.

Consider a simple proof of the theorem. Let the first qubit be in the state $|\psi\rangle$, and the second qubit be in some state $|s\rangle$. Let us assume that by means of a certain unitary transformation U it is possible to copy the desirable state: $|\psi\rangle \otimes |s\rangle \rightarrow U(|\psi\rangle \otimes |s\rangle) = |\psi\rangle \otimes |\psi\rangle$. Next, let us perform the same copying procedure with the state $|\phi\rangle$. Then we have $U |\psi\rangle |s\rangle = |\psi\rangle |\psi\rangle$ and $U |\phi\rangle |s\rangle = |\phi\rangle |\phi\rangle$. We take the inner product of the two equalities. This means that we multiply the former with the conjugate of the latter, $\langle s| \langle \phi| U^\dagger = \langle \phi| \langle \phi|$, and we obtain $\langle \psi |\phi\rangle = \langle \psi |\phi\rangle^2$. This equation has the form $x = x^2$, and its

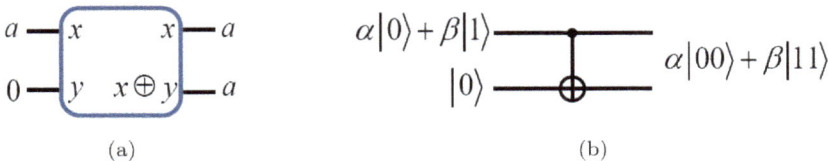

(a)   (b)

Fig. 1.7.   Classical circuit for copy operation (a) and its quantum analogue (b).

solution is either 0 or 1. So, we can clone only orthogonal (basic) states, and not an arbitrary state.

1.5.3. *Bell states*

In quantum information it is important to consider the so-called Bell states. These can be introduced as the states derived from the basis states by making use of the Hadamard operation and the controlled-NOT operation, as shown in Fig. 1.8.

Then we obtain:

$$|00\rangle \rightarrow \frac{|00\rangle + |11\rangle}{\sqrt{2}} \equiv |B_{00}\rangle, \quad |01\rangle \rightarrow \frac{|01\rangle + |10\rangle}{\sqrt{2}} \equiv |B_{01}\rangle,$$

$$|10\rangle \rightarrow \frac{|00\rangle - |11\rangle}{\sqrt{2}} \equiv |B_{10}\rangle, \quad |11\rangle \rightarrow \frac{|01\rangle - |10\rangle}{\sqrt{2}} \equiv |B_{11}\rangle. \tag{1.22}$$

These states are maximally entangled. This means that they are significantly far from the separable states of the form $|00\rangle + |01\rangle = |0\rangle (|0\rangle + |1\rangle)$. The *entangled states* are defined as the states which cannot be rewritten in the form of a tensor product of one-particle states. The states above are called the *Bell states* or *EPR pairs*. Indeed, consider a separable state

$$(\alpha_1 |0\rangle + \beta_1 |1\rangle)(\alpha_2 |0\rangle + \beta_2 |1\rangle)$$

$$= \alpha_1\alpha_2 |00\rangle + \alpha_1\beta_2 |01\rangle + \beta_1\alpha_2 |10\rangle + \beta_1\beta_2 |11\rangle. \tag{1.23}$$

From this, to obtain, say, $|B_{00}\rangle$, it is necessary to make the factors $\alpha_1\beta_2$ and $\beta_1\alpha_2$ zero, but then we cannot obtain nontrivial factors before $|00\rangle$ and $|11\rangle$.

Here EPR stands for the names Einstein, Podolsky, and Rosen, who, in relation to the interpretation of such states, formulated the known paradox (see Footnote 3 on page 3). Their paradox

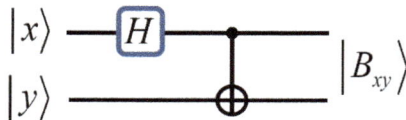

Fig. 1.8.   Preparation of the Bell states $|B_{xy}\rangle$.

expresses the hesitation in the existence of quantum correlations. However, since the existence of these correlations has been proven, it is probably correct to refer to this as the effect rather than the paradox. In this relation it is "astonishing that in standard textbooks of quantum mechanics the notion of the entanglement in quantum systems is often even not mentioned" [Valiev 2005].

So, consider the EPR paradox in its simplified formulation. Let the pair of quantum particles be prepared in an entangled state, say, in the singlet state $|B_{11}\rangle$. Assume that these particles are separated in space. The particles continue to stay in the entangled state with the total spin being 0. Because of the quantum correlation of these two particles (which are in the entangled state), the measurement of one of them immediately defines the value of another — in our case, with the opposite spin. Namely, if the measurement of the first particle state gives $|0\rangle$, then the second particle is in the state $|1\rangle$; and if for the first particle we have $|1\rangle$, then for the second one we know that its state is $|0\rangle$. But there is no contradiction, if one takes into account the *nonlocality of quantum correlations*, which explains this "immediate long-range interaction". Note that the information about the measurement should be transferred via a classical channel and there is no contradiction that the information is travelling faster than the light velocity (which was assumed by the EPR paradox). EPR claimed that quantum mechanics is not complete and there are "hidden parameters" which define the result of a measurement and do not contradict the physical principle of locality. Later this dilemma was formulated in the form of the so-called Bell inequalities. Convincing experimental proofs of the Bell inequalities gave the answer to these questions: *the quantum mechanics is complete, it does not contain hidden parameters, and the quantum correlations bear nonlocal character.*

Now, before we consider the Bell inequalities, in addition to what was said, it is appropriate to review how measurements are described.

1.5.4. *Projective measurements*

Measurements are described by the measurement operators $\{M_m\}$ (they are also called the projectors on the proper subspace of the operator M). If a system is in the state $|\psi\rangle$, then the probability of measuring the result m equals $p(m) = \langle\psi| M_m^\dagger M_m |\psi\rangle$, and after the measurement, the system can be found in the state $M_m |\psi\rangle / \sqrt{p(m)}$. (At that, the completeness condition is fulfilled, $\sum M_m^\dagger M_m = I$, which means that $\sum p(m) = 1$.)

In particular, for one qubit in the calculation basis, that is in the basis $\{|0\rangle, |1\rangle\}$, we have $M_m = |m\rangle\langle m|$, and there are relations $M_m^\dagger = M_m$ and $M_m^2 = M_m$. For example, consider the state $|\psi\rangle = \alpha|0\rangle + \beta|1\rangle$ and the measurement result of the state $|0\rangle$ with the operator $M_0 = |0\rangle\langle 0|$; this appears with the probability $p(0) = |\alpha|^2$, and the state after the measurement will be $\alpha|0\rangle / |\alpha|$.

Consider now the probability of measuring the first qubit in the $|0\rangle$ state for a system of two qubits in the state $|B_{11}\rangle$: with the operator $M_0^{(1)} = \sum |0j\rangle\langle 0j| = |00\rangle\langle 00| + |01\rangle\langle 01|$ we obtain $p^{(1)}(0) = \langle M_0^{(1)}\rangle = \langle B_{11}|M_0^{(1)}|B_{11}\rangle = 1/2$ and the wave function after the measurement $M_0^{(1)}|B_{11}\rangle / \sqrt{p^{(1)}(0)} = |01\rangle \equiv |0\rangle|1\rangle$. For an arbitrary state $|\psi\rangle = \alpha_{00}|00\rangle + \alpha_{01}|01\rangle + \alpha_{10}|10\rangle + \alpha_{11}|11\rangle$ the probability of the first qubit to be in the state $|0\rangle$ equals $p^{(1)}(0) = |\alpha_{00}|^2 + |\alpha_{01}|^2$, and after the measurement we obtain $|\psi'\rangle = (\alpha_{00}|00\rangle + \alpha_{01}|01\rangle) / \sqrt{p^{(1)}(0)}$.

An arbitrary projective measurement is described by the operator $M = \sum m M_m = \sum m |m\rangle\langle m|$. Then for the expectation value we have $E(M) = \sum mp(m) = \sum m \langle\psi| M_m |\psi\rangle = \langle\psi| \sum m M_m |\psi\rangle = \langle\psi| M |\psi\rangle$. For example, let the problem be formulated to define one of the Bell states. We enumerate them by the number m, from 1 to 4, so that $\{|m\rangle\} = \{|B_{00}\rangle, |B_{01}\rangle, |B_{10}\rangle, |B_{11}\rangle\}$. Then the projective measurement of a certain state $|k\rangle$ gives

$$M = \langle M\rangle = \langle k| \left(\sum_m m |m\rangle\langle m|\right) |k\rangle = \sum_m m\delta_{mk} = k. \qquad (1.24)$$

In conclusion of this excursus into quantum mechanics, we note that the measurement of one of the basis states does not change this state. This means that if we measure a basis state $|k\rangle$ in this basis, then we obtain $p(m) = \langle k| M_m^\dagger M_m |k\rangle = \delta_{mk}$, which in turn means that with the probability $p(k) = 1$ we find the system in the state $|k\rangle$.

At the same time, after the measurement, the system can be found in the state $M_m |k\rangle / \sqrt{p(m)} = |k\rangle$. Hence, such measurement does not change the system state and is called the *quantum non-demolition measurement* (QND).

1.5.5. *Bell inequalities*

Let us compare correlations in classical and quantum cases using the example of the measurement series of a two-qubit system.[12,13] Let the system be initially prepared in the singlet state $|B_{11}\rangle = \frac{|01\rangle - |10\rangle}{\sqrt{2}}$. The measurement of a spin in a certain direction of a vector $\vec{a}$ is given by the projection of the spin operator $\vec{\sigma} = (\sigma_1, \sigma_2, \sigma_3)$ on this selected direction, $\sigma_a = \sigma_i a_i$; the eigenvalues of the spin-projection operator are ± 1. Let the measurable value be the projection of the first spin on $\vec{a}$ and the projection of the second spin be on $\vec{b}$. One can demonstrate the following:

$$\langle B_{11}| \vec{\sigma}\vec{a} \otimes \vec{\sigma}\vec{b} |B_{11}\rangle = -\vec{a}\vec{b}. \qquad (1.25)$$

This can be done, for example, by considering this correlator component-wise:

$$\langle B_{11}| \vec{\sigma}\vec{a} \otimes \vec{\sigma}\vec{b} |B_{11}\rangle = a_i b_j \langle B_{11}| \sigma_i^{(1)} \sigma_j^{(2)} |B_{11}\rangle$$

$$= a_i b_j \frac{1}{2} \left(\sigma_{i,00}^{(1)} \sigma_{j,11}^{(2)} + \cdots \right) = -a_i b_i. \qquad (1.26)$$

Here $\sigma_i^{(1)} = \sigma_i \otimes \sigma_0$ and $\sigma_i^{(2)} = \sigma_0 \otimes \sigma_i$ are the spin matrices corresponding to the operators of the first and the second qubits; σ_0 is the unity matrix. Indeed,

$$\vec{\sigma}\vec{a} \otimes \vec{\sigma}\vec{b} = \vec{\sigma}\vec{a} \cdot \sigma_0 \otimes \sigma_0 \cdot \vec{\sigma}\vec{b} = \vec{\sigma}^{(1)}\vec{a} \cdot \vec{\sigma}^{(2)}\vec{b}, \qquad (1.27)$$

where it was taken into account that $A \otimes B \cdot C \otimes D = AC \otimes BD$.

---

[12] A. S. Holevo, Introduction to the quantum theory of information: Moscow, lectures in the Russian Quantum Center (2013).
[13] N. V. Evdokimov, D. N. Klyshko, V. P. Komolov, V. A. Yarochkin, Bell's inequalities and EPR-Bohm correlations: working classical radiofrequency model, Phys.–Uspekhi **39**, 83 (1996).

Let us define the so-called Bell observable as a result of the four measurements:

$$S_q \equiv \left\langle \sigma_{a_1}^{(1)} \sigma_{b_1}^{(2)} \right\rangle + \left\langle \sigma_{a_1}^{(1)} \sigma_{b_2}^{(2)} \right\rangle + \left\langle \sigma_{a_2}^{(1)} \sigma_{b_1}^{(2)} \right\rangle - \left\langle \sigma_{a_2}^{(1)} \sigma_{b_2}^{(2)} \right\rangle$$
$$= -\mathbf{a_1 b_1} - \mathbf{a_1 b_2} - \mathbf{a_2 b_1} + \mathbf{a_2 b_2}. \tag{1.28}$$

Consider now the correlations in the classical model that satisfy the locality principle. An analogous Bell observable for the random values X_i and Y_i, such that $|X_i|, |Y_i| \leq 1$, will be defined as

$$S_{cl} \equiv E(X_1 Y_1) + E(X_1 Y_2) + E(X_2 Y_1) - E(X_2 Y_2). \tag{1.29}$$

Here $E(X) = \sum_m x_m p_m$ is the expectation of the value X, defined by arbitrary probabilities p_m. The Bell inequality can be written in the form

$$S_{cl} \leq 2, \tag{1.30}$$

which is known as the *Clauser–Horne–Shimony–Holt (CHSH) inequality*. This follows from averaging the inequality

$$X_1 Y_1 + X_1 Y_2 + X_2 Y_1 - X_2 Y_2 \leq 2. \tag{1.31}$$

Here one can speculate more simply: let $X_i = Y_i = \pm 1$. Then

$$X_1 Y_1 + X_1 Y_2 + X_2 Y_1 - X_2 Y_2$$
$$= X_1(Y_1 + Y_2) + X_2(Y_1 - Y_2) = \pm 2, \tag{1.32}$$

and averaging this equality gives Eq. (1.30).

We return to the quantum correlations. Let us point the axes in one plane with the following polar angles: 0 and $\pi/2$ for $\mathbf{a}_{1,2}$ and $5\pi/4$ and $3\pi/4$ for $\mathbf{b}_{1,2}$. We obtain, for the observable defined by Eq. (1.27):

$$S_q = 2\sqrt{2}. \tag{1.33}$$

So, classical description for the correlations of the spins of two particles does not allow such value of correlation to be reached, which corresponds to the predictions of quantum mechanics. And as was pointed out above, multiple experimental verifications have confirmed this latter statement: they observed the correlations, corresponding to the predictions of the quantum mechanics, assuming violation of the Bell inequalities and non-local character of the correlations; see, for example, the beautiful recent work in Footnote 14.

1.5.6. *Superdense coding*

To illustrate a use of quantum mechanics, consider first the problem on superdense coding [Nielsen and Chuang 2010]. In the simplest approach, the problem is formulated as the transference of two bits of information from, as they say, Alice to Bob. Quantum mechanics allows this by transmitting only one qubit.

Consider Alice and Bob having, in the initial moment of time, one qubit each out of the Bell pair $|B_{00}\rangle = \frac{|00\rangle + |11\rangle}{\sqrt{2}}$. To solve the formulated problem of transferring two bits of information, (i, j), Alice makes the one-qubit operation with her qubit, which transfers the qubit-pair state into the corresponding Bell state. Namely, if she wants to transfer $ij = 00$, then she does not do anything (makes the unitary transformation), and further

$$01: |B_{00}\rangle \xrightarrow{X} \frac{|10\rangle + |01\rangle}{\sqrt{2}} = |B_{01}\rangle,$$

$$10: |B_{00}\rangle \xrightarrow{Z} \frac{|00\rangle - |11\rangle}{\sqrt{2}} = |B_{10}\rangle, \qquad (1.34)$$

$$11: |B_{00}\rangle \xrightarrow{iY} \frac{-|10\rangle + |01\rangle}{\sqrt{2}} = |B_{11}\rangle.$$

Namely, this means that to transfer the bits $(1,0)$, Alice makes the Z operation with her qubit and so forth. As the result of the respective operation, Alice changes the state of the pair: $|B_{00}\rangle \rightarrow |B_{ij}\rangle$. After this, she sends her one qubit to Bob. Now, Bob has the whole EPR pair. Making the measurement in the EPR basis, Bob gets the state $|B_{ij}\rangle$ with the unitary probability, as we discussed after Eq. (1.24). So, after Alice transmits her *one* qubit, Bob receives *two* bits of information.

---

[14]J. Handsteiner *et al.*, Phys. Rev. Lett. **118**, 060401 (2017).

We note that to measure the Bell states, one can first change them to the calculation-basis states. For this, Bob should dis-entangle the Bell state by the circuit inverse to the one in Fig. 1.8. Making CNOT and H_1 gates, he would transform $|B_{00}\rangle \to |00\rangle$, $|B_{01}\rangle \to |01\rangle$, $|B_{10}\rangle \to |10\rangle$, $|B_{11}\rangle \to |11\rangle$.

Moreover, such a way of transferring information is highly protected. Indeed, assume a malefactor intercepts a qubit sent by Alice. She (let this be a woman) measures this qubit with some operator M. But having only one qubit, she gets an identical result for the four Bell states. Say, for $|B_{10}\rangle$ she obtains:

$$\langle B_{10}| M \otimes I |B_{10}\rangle = \frac{1}{2}((\langle 00| - \langle 11|)M \otimes I(|00\rangle - |11\rangle)$$

$$= \frac{1}{2}(M_{00} + M_{11}). \tag{1.35}$$

And, as it is easy to check, she would obtain the very same result for other $|B_{ij}\rangle$ states. Here it is taken into account that the operator M acts only on the first qubit, and that is why for the basis states we have

$$\langle i_2 i_1| M |j_1 j_2\rangle = \langle i_2| \langle i_1| M |j_1\rangle |j_2\rangle = M_{i_1 j_1} \delta_{i_2 j_2}. \tag{1.36}$$

1.5.7. *Quantum teleportation*

Consider transferring *quantum* information from Alice to Bob. Assume they have one part from the EPR pair each. After this, Alice, via classical channel, tells Bob the key, and he immediately knows about the state of Alice's qubit. So, we aim to consider here the principle of quantum teleportation, which is the technique of transferring quantum information in the absence of a quantum information channel [Nielsen and Chuang 2010].

To solve this problem, Alice puts her principal qubit, of which the state $|\psi\rangle$ needs to be transferred, into contact with her half of the EPR pair, see Fig. 1.9. And then she measures the state of her two qubits. She sends the information about her qubits' states to Bob. With this information Bob restores the state $|\psi\rangle$, using his half of the EPR pair. Such a solution of the problem is illustrated in Fig. 1.9; and this is described as follows.

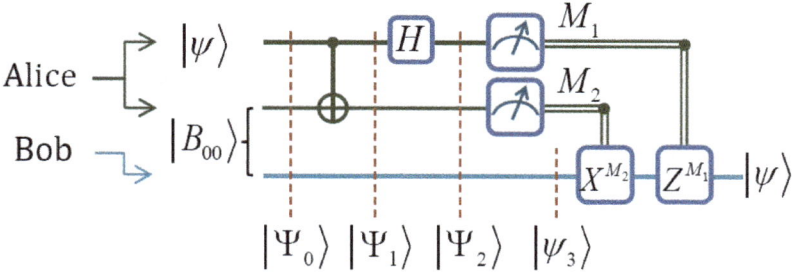

Fig. 1.9.   Schematic of the qubit quantum teleportation.

Well, we need to transfer the state $|\psi\rangle = \alpha |0\rangle + \beta |1\rangle$. At the input we have

$$|\Psi_0\rangle = |\psi\rangle |B_{00}\rangle = \frac{1}{\sqrt{2}}[\alpha |0\rangle (|00\rangle + |11\rangle) + \beta |1\rangle (|00\rangle + |11\rangle)],$$
(1.37)

where the first two qubits are at Alice's disposal and the latter qubit is what Bob has. Alice passes her pair through the controlled-NOT gate,

$$|\Psi_1\rangle = \frac{1}{\sqrt{2}}[\alpha |0\rangle (|00\rangle + |11\rangle) + \beta |1\rangle (|10\rangle + |01\rangle)].$$
(1.38)

Then she applies the Hadamard operation to the first qubit:

$$|\Psi_2\rangle = \frac{1}{2}[\alpha (|0\rangle + |1\rangle)(|00\rangle + |11\rangle) + \beta (|0\rangle - |1\rangle)(|10\rangle + |01\rangle)]$$

$$= \frac{1}{2}[|00\rangle (\alpha |0\rangle + \beta |1\rangle) + |01\rangle (\alpha |1\rangle + \beta |0\rangle)$$

$$+ |10\rangle (\alpha |0\rangle - \beta |1\rangle) + |11\rangle (\alpha |1\rangle - \beta |0\rangle)].$$
(1.39)

It is worth reminding that we assume such notations: $|0\rangle |10\rangle \equiv |010\rangle \equiv |0_1 1_2 0_3\rangle \equiv |0_1 1_2\rangle |0_3\rangle \equiv |01\rangle |0\rangle$. We can see from here that as soon as Alice makes the measurement, Bob's qubit is reduced to one of the four states:

$$00 \Rightarrow |\psi_3(00)\rangle \equiv [\alpha |0\rangle + \beta |1\rangle],$$
$$01 \Rightarrow |\psi_3(01)\rangle \equiv [\alpha |1\rangle + \beta |0\rangle], \ \dots.$$
(1.40)

After this Bob only needs to correct his result. In the first case, he does not even have to do anything. In the second case he has to make the X operation, and so forth. To be more precise, Bob has to make the operation $Z^{M_1} X^{M_2}$, where M_1 and M_2 stand for the values measured by Alice on the first and second qubits.

Moreover, the quantum teleportation algorithm, considered here, demonstrates the *interchangeability of diverse resources* in the quantum mechanics: here, working with an EPR pair and transmission of two classical bits of information was equivalent to the transmission of one qubit of information.

1.5.8. *Quantum parallelism — Deutsch algorithm*

What differentiates a quantum computer from a classical computer, in essence, is quantum parallelism. This assumes the ability to simultaneously calculate a binary function $f(x)$ for diverse values x. For instance, assume the transformation U_f transfers the state $|x, y\rangle$ into $|x, y \oplus f(x)\rangle$. Then, if we prepare the first qubit in the state $|x\rangle = (|0\rangle + |1\rangle)/\sqrt{2}$ and the second qubit in the state $|y\rangle = |0\rangle$, the operation U_f gives, at the output, the state $(|0, f(0)\rangle + |1, f(1)\rangle)/\sqrt{2}$. This procedure is a prototype of the Deutsch algorithm. This allows the two values of the function, $f(0)$ and $f(1)$, to be probed with one measurement. To be more precise, we can measure only one result, for example, $f(0) \oplus f(1)$. Thus, using the state superposition, quantum parallelism can be realized as an execution of the simultaneous calculation of n values of the function $f(x)$ on n qubits. Note that for this, a classical computer needs n calculations.

Consider, for illustrative purposes, the problem of "how to see two sides of a coin simultaneously." Assume a binary variable $x = (0, 1)$ defines the coin side, and the function $f(x) = (0, 1)$ defines the heads or tails. If we define $f(x)$ twice for the two values of x, we can get four options for the answer: 0, 1, x, NOT(x). In the first two cases the coin is false; in the last two cases the coin is not false. For definition of the authenticity of a coin in a classical case, obviously, we need two measurements, while in the quantum case one measurement suffices.

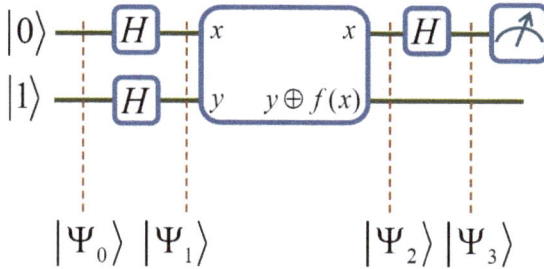

Fig. 1.10.   The algorithm, which realizes quantum parallelism. This demonstrates solution of the problem, "how to see two sides of a coin simultaneously."

The method of achieving this is demonstrated as a simple Deutsch algorithm in Fig. 1.10.

At the input we feed the state $|\Psi_0\rangle = |01\rangle \equiv |0\rangle |1\rangle$. The Hadamard operation changes this state to the following

$$|\Psi_1\rangle = \frac{|0\rangle + |1\rangle}{\sqrt{2}} \frac{|0\rangle - |1\rangle}{\sqrt{2}}. \tag{1.41}$$

Next, let us make the operation U_f such that $|x, y\rangle \rightarrow |x, y \oplus f(x)\rangle$. It can be demonstrated that this operation works as a phase transformation:

$$|x\rangle (|0\rangle - |1\rangle) \rightarrow |x\rangle (|f(x)\rangle - |1 \oplus f(x)\rangle)$$
$$= (-1)^{f(x)} |x\rangle (|0\rangle - |1\rangle). \tag{1.42}$$

Here it was taken into account that $|f(x)\rangle - |1 \oplus f(x)\rangle = |0\rangle - |1\rangle$ for $f(x) = 0$ and $|f(x)\rangle - |1 \oplus f(x)\rangle = |1\rangle - |0\rangle$ for $f(x) = 1$. Then we obtain

$$|\Psi_2\rangle = \frac{1}{2} \left((-1)^{f(0)} |0\rangle + (-1)^{f(1)} |1\rangle \right) (|0\rangle - |1\rangle) \tag{1.43}$$

or

$$|\Psi_2\rangle = \begin{cases} \pm \dfrac{|0\rangle + |1\rangle}{\sqrt{2}} \dfrac{|0\rangle - |1\rangle}{\sqrt{2}}, & f(0) = f(1), \\[4mm] \pm \dfrac{|0\rangle - |1\rangle}{\sqrt{2}} \dfrac{|0\rangle - |1\rangle}{\sqrt{2}}, & f(0) \neq f(1). \end{cases} \tag{1.44}$$

A subsequent Hadamard operation on the first qubit gives

$$|\Psi_3\rangle = \begin{cases} \pm|0\rangle \dfrac{|0\rangle - |1\rangle}{\sqrt{2}}, & f(0) = f(1), \\[3mm] \pm|1\rangle \dfrac{|0\rangle - |1\rangle}{\sqrt{2}}, & f(0) \neq f(1). \end{cases} \tag{1.45}$$

This is equivalent to the following

$$|\Psi_3\rangle = \pm|f(0) \oplus f(1)\rangle \frac{|0\rangle - |1\rangle}{\sqrt{2}}, \tag{1.46}$$

if we take into account that $f(0) \oplus f(1) = 0$ for $f(0) = f(1)$ and 1 for $f(0) \neq f(1)$. So, the measurement of the first qubit's state gives the answer for the formulated question:

$$\begin{aligned} f(0) \oplus f(1) = 0 &\Rightarrow f(x) = const, \\ f(0) \oplus f(1) = 1 &\Rightarrow f(x) \neq const. \end{aligned} \tag{1.47}$$

And thus, this algorithm allows definition of the global property of the function after one measurement, namely to calculate $f(0) \oplus f(1)$, which requires one run here, instead of two runs as in the classical case.

The problem considered demonstrates pictorially the advantages of quantum correlations (entanglement), which do not reduce to classical correlations. Also, one can consider here other examples, demonstrating "quantum supremacy", including such beautiful problems as the Merlin–Peres quantum telepathic game (§2.5 in Footnote 12, page 27 of this book) and the Elitzur–Vaidman problem about testing bombs (Chapter 2 in Footnote 15 below).

Conclusion to Chapter 1

In this chapter we pursued mainly two aims: to learn about quantum information and to discuss related ideas in quantum mechanics, which are both fundamental and important for applications.

We became familiar with ideas such as a qubit, quantum gates and algorithms, the no-cloning theorem, quantum teleportation and

---

[15]R. Penrose, A. Shimony, N. Cartwright, S. Hawking, The Large, the Small and the Human Mind, Cambridge, UK: Cambridge University Press (2000).

parallelism. On the other hand, we have considered microscopic realizations of qubits, such as spins and photons, as well as discussed the superposition and entanglement of quantum states, Bell inequality and other basic concepts in quantum mechanics.

In conclusion, it is appropriate to note that all these follow from only several postulates, which describe the relation between the physical world and the mathematical language of quantum mechanics. That is why we will lay them out here, separately. Note that in this Chapter, we have followed the textbook [Nielsen and Chuang 2010] to a large degree.

Postulate 1. An isolated system is described fully by the vector-state, which is defined in the space of the system states. In this space the principle of superposition is satisfied.

Postulate 2. The time evolution of the state of a closed quantum system is described by the Schrödinger equation $i\hbar\frac{d}{dt}|\psi\rangle = H|\psi\rangle$. The solution of this equation is $|\psi(t)\rangle = \exp\left(-\frac{iH}{\hbar}t\right)|\psi(0)\rangle$; this means that the evolution is described by a unitary transformation.

Postulate 3. Quantum measurements are described by a set of measurement operators $\{M_m\}$. If before the measurement, the system was in a state $|\psi\rangle$, then the probability of getting the result m is $p(m) = \langle\psi|M_m^\dagger M_m|\psi\rangle$; after the measurement, the system will be in the state $M_m|\psi\rangle/\sqrt{p(m)}$.

Postulate 4. The space of states of a composite system is formed by the tensor product of the subsystem subspaces. This postulate can be interpreted as a generalization of the superposition principle for the description of a composite system.

After reading this Chapter, we can see that Quantum Information theory is largely based on the basic postulates of Quantum Mechanics. On the other hand, we can see how the fundamental notions of Quantum Mechanics become the working tools and language of Quantum technologies.

Problems for independent work and for self-assessment

1.1. (*) Describe the action of the basic one-qubit operations, Eqs. (1.13)–(1.14).

1.2. (*) Obtain the matrix for the CNOT operation.

1.3. (*) Describe the swap operation.

1.4. (**) Prove the no-cloning theorem.

1.5. (*) Making use of the Hadamard and CNOT operations, obtain the Bell states.

1.6. (**) Prove the relation (1.25), describing the measurement of the two-spin system.

1.7. (***) Given the algorithm in Fig. 1.9, describe quantum teleportation.

1.8. (***) Given the algorithm in Fig. 1.10, describe quantum parallelism.

Chapter 2

QUANTUM MECHANICS OF QUBITS

(Dynamical behaviour of a two-level system)

> "In addition to the well-studied statics of the quantum systems, mesoscopic physics adds the new aspects of dynamics in quantum mechanics."
>
> [Valiev 2005]

The problem of dynamical behaviour in a two-level system deserves detailed discussion. First, this gives results for fundamental problems. Second, it is very topical for mesoscopic systems, which have parameters tunable in a wide range, and where the regimes of control and interferometry are important. Third, it is a good example of the accurate solution of a realistic problem. Fourth, this will allow us to introduce useful formulas and approaches.

2.1. Two-level system

Consider a two-level system driven periodically. A two-level system with the energy bias ε and the tunneling amplitude Δ is described by the pseudospin Hamiltonian

$$H(t) = -\frac{\Delta}{2}\sigma_x - \frac{\varepsilon(t)}{2}\sigma_z \qquad (2.1)$$

in terms of the Pauli matrices $\sigma_{x,z}$ (we wrote about this above, when we discussed Eq. (1.9)). Usually, the value Δ is assumed to be constant, while the bias ε is considered to be a time-dependent controlling parameter. The most interesting is the situation with monochromatic time dependence,

$$\varepsilon(t) = \varepsilon_0 + A\cos\omega t, \tag{2.2}$$

with the amplitude A, frequency ω, and offset ε_0.

We can split the one-qubit Hamiltonian into the time-independent and time-dependent parts, $H = H_0 + V(t)$ with

$$H_0 = -\frac{\Delta}{2}\sigma_x - \frac{\varepsilon_0}{2}\sigma_z, \quad V(t) = -\frac{A\cos\omega t}{2}\sigma_z. \tag{2.3}$$

First, let us define eigenvectors and eigenfunctions of the operator H_0 from the stationary Schrödinger equation $H_0|\psi\rangle = E|\psi\rangle$. Then for $|\psi\rangle = \alpha|0\rangle + \beta|1\rangle$ we have

$$(H_0 - E)|\psi\rangle = -\frac{1}{2}\begin{pmatrix} \varepsilon_0 + 2E & \Delta \\ \Delta & -\varepsilon_0 + 2E \end{pmatrix}\begin{pmatrix} \alpha \\ \beta \end{pmatrix} = 0. \tag{2.4}$$

Equating the determinant to zero, we obtain

$$E = E_\pm \equiv \pm\frac{1}{2}\sqrt{\Delta^2 + \varepsilon_0^2} \equiv \pm\frac{1}{2}\Delta E, \tag{2.5}$$

where we defined the distance between the *qubit energy levels* $\Delta E = E_+ - E_- = \sqrt{\Delta^2 + \varepsilon_0^2}$.

Further we solve the system of equations (2.4); from the first equation it follows

$$\alpha = -\frac{\Delta}{\varepsilon_0 + 2E}\beta, \tag{2.6}$$

and from the normalization condition $1 = \alpha^2 + \beta^2$ we have

$$\beta^2 = \frac{1}{1 + \frac{\Delta^2}{(\varepsilon_0 + 2E)^2}} = \frac{(\varepsilon_0 \pm \Delta E)^2}{(\varepsilon_0 \pm \Delta E)^2 + \Delta^2} = \frac{(\Delta E \pm \varepsilon_0)^2}{2\Delta E(\Delta E \pm \varepsilon_0)}$$

$$= \frac{1}{2}\left(1 \pm \frac{\varepsilon_0}{\Delta E}\right) \equiv \gamma_\pm^2. \tag{2.7}$$

Then we find α from the expression $\alpha^2 = 1 - \beta^2$ and take into account that from Eq. (2.6) we have sgn $\alpha = \mp$sgn β. So, $\alpha_\pm = \gamma_\mp$ and $\beta_\pm = \mp\gamma_\pm$, and for the eigenfunctions of H_0 we have

$$|E_-\rangle = \gamma_+|0\rangle + \gamma_-|1\rangle,$$
$$|E_+\rangle = \gamma_-|0\rangle - \gamma_+|1\rangle, \tag{2.8}$$
$$\gamma_\pm = \frac{1}{\sqrt{2}}\sqrt{1 \pm \frac{\varepsilon_0}{\Delta E}}.$$

This means, in particular, that at the point where the levels maximally approach (or, as they say, at the point of the levels quasicrossing), where $\varepsilon_0 = 0$, we have

$$|E_\pm\rangle = \frac{|0\rangle \pm |1\rangle}{\sqrt{2}}, \tag{2.9}$$

while far from this point: $|E_-\rangle = |1\rangle$ and $|E_+\rangle = |0\rangle$ at $\varepsilon_0 \to -\infty$ and $|E_-\rangle = |0\rangle$ and $|E_+\rangle = -|1\rangle$ at $\varepsilon_0 \to \infty$. In Fig. 2.1 the levels of the ground $|E_-\rangle$ and excited $|E_+\rangle$ states are shown by the solid lines, and the dashed lines show the so-called *diabatic* states. These states are defined as the eigenstates of the Hamiltonian with $\Delta = 0$, i.e. $H_0 = -\frac{\varepsilon_0}{2}\sigma_z$; they are $E_{\uparrow,\downarrow} = \mp\frac{\varepsilon_0}{2}$.

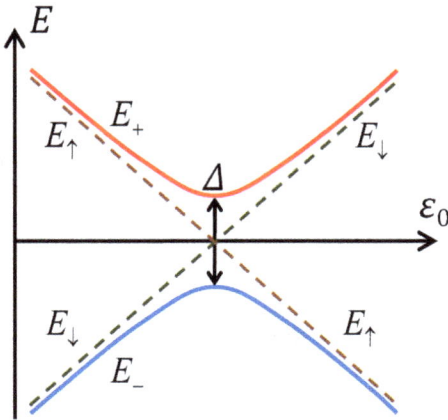

Fig. 2.1.   Qubit energy levels with dependence on the energy bias ε_0.

It is useful also to arrive at the basis of the eigenstates $|E_\pm\rangle$ in a different way. For this, we note that the effective magnetic field in the Hamiltonian (2.1) lies in the xz plane and we want to have this along the z axis. For this we need to make the rotation around the y axis:

$$S = \exp\frac{i\zeta\sigma_y}{2} = \cos\frac{\zeta}{2} + i\sigma_y\sin\frac{\zeta}{2}, \qquad (2.10)$$

where it was taken into account that $\exp i\alpha\vec{n}\vec{\sigma} = \cos\alpha + i\vec{n}\vec{\sigma}\sin\alpha$. Let us find the rotation angle ζ. Note that the unitary transformation S changes the vector-state, $|\psi\rangle = S|\psi'\rangle$; at this, the Schrödinger equation $i\hbar\frac{\partial}{\partial t}|\psi\rangle = H|\psi\rangle$ takes the form $i\hbar S\frac{\partial}{\partial t}|\psi'\rangle = HS|\psi'\rangle$ or $i\hbar\frac{\partial}{\partial t}|\psi'\rangle = H'|\psi'\rangle$, where the new Hamiltonian is $H' = S^\dagger HS$. We would like to have this diagonal,

$$H' = -\frac{\Delta E}{2}\sigma_z \equiv -\frac{\Delta E}{2}\left(|E_-\rangle\langle E_-| - |E_+\rangle\langle E_+|\right), \qquad (2.11)$$

so that

$$H'|E_-\rangle = -\frac{\Delta E}{2}|E_-\rangle \equiv E_-|E_-\rangle,$$

$$H'|E_+\rangle = \frac{\Delta E}{2}|E_+\rangle \equiv E_+|E_+\rangle. \qquad (2.12)$$

We obtain the equation for the desired rotation angle

$$H = -\frac{1}{2}\begin{pmatrix} \varepsilon_0 & \Delta \\ \Delta & -\varepsilon_0 \end{pmatrix} = SH'S^\dagger. \qquad (2.13)$$

From here, by multiplying the matrices in the r.h.s., we obtain that ζ is defined by the following: $\sin\zeta = -\Delta/\Delta E$ and $\cos\zeta = \varepsilon_0/\Delta E$; that is, $\tan\zeta = -\Delta/\varepsilon_0$. In particular, for the relation between the basis vectors, we can write $|E_i\rangle = S_{ij}^\dagger|m_j\rangle$, where $|E_i\rangle = \{|E_-\rangle, |E_+\rangle\}$ and $|m_j\rangle = \{|0\rangle, |1\rangle\}$. This means

$$|E_-\rangle = \cos\frac{\zeta}{2}|0\rangle + \sin\frac{\zeta}{2}|1\rangle,$$

$$|E_+\rangle = \sin\frac{\zeta}{2}|0\rangle - \cos\frac{\zeta}{2}|1\rangle, \qquad (2.14)$$

which coincide with the formulas (2.8).

2.2. Rabi oscillations

Even though a two-level system is one of the basic systems of physics, specific calculations may present difficulties. In particular, the Schrödinger equation for a two-level system with periodic driving can be written in the form of the second-order differential equation with periodic coefficients, which is the Hill equation. This cannot be solved by quadratures. Nevertheless, in different regions of the parameter space, we can use approximate analytic results. The choice of approach depends on a specific problem and on the interrelation among three key parameters: A, Δ, and ω.

Consider, first, a two-level system subjected to a weak periodic field, which is described by the Hamiltonian (2.1). In this subsection we will consider the excitation with weak amplitude and with the frequency ω close to the resonant qubit frequency $\omega_q \equiv \Delta E / \hbar$:

$$A \ll \Delta, \quad \delta\omega \equiv \omega - \omega_q \ll \omega. \tag{2.15}$$

Let us first switch to the representation of the eigenstates of H_0 with the help of the transformation S described above. Then we get the Hamiltonian

$$H' = S^\dagger (H_0 + V(t)) S = -\frac{\Delta E}{2} \sigma_z - \frac{A \cos \omega t}{2} (\cos \zeta \cdot \sigma_z + \sin \zeta \cdot \sigma_x). \tag{2.16}$$

We remind ourselves that in the new representation, the eigenstates of H_0 have the form:

$$|E_-\rangle = \begin{pmatrix} 1 \\ 0 \end{pmatrix}, \quad |E_+\rangle = \begin{pmatrix} 0 \\ 1 \end{pmatrix}. \tag{2.17}$$

This means that they are the eigenstates of the operator σ_z.

For the wave function we choose the following ansatz

$$|\psi'\rangle = a(t) e^{-i\frac{E_-}{\hbar}t} |E_-\rangle + b(t) e^{-i\frac{E_+}{\hbar}t} |E_+\rangle = \begin{pmatrix} a e^{i\frac{\omega_q}{2}t} \\ b e^{-i\frac{\omega_q}{2}t} \end{pmatrix}, \tag{2.18}$$

which means that we expanded the wave function with the eigen-energy states depending on time. Then from the Schrödinger equation $i\hbar\frac{\partial}{\partial t}|\psi'\rangle = H'|\psi'\rangle$ we obtain

$$i\hbar\begin{pmatrix} \dot{a} \\ \dot{b} \end{pmatrix} = -\frac{A}{4\Delta E}(e^{i\omega t} + e^{-i\omega t})\begin{pmatrix} \varepsilon_0 a - \Delta b e^{-i\omega_q t} \\ -\Delta a e^{i\omega_q t} - \varepsilon_0 b \end{pmatrix}. \tag{2.19}$$

Since we consider the excitation close to the resonance, $\omega \sim \omega_q$, we can omit all the "fast-rotating" terms. This means that we can omit the terms with the fast time dependence (for instance, $e^{\pm i\omega t}$) and leave the slowly varying terms of the form $e^{\pm i\delta\omega t}$. This procedure can also be explained by averaging Eq. (2.19) over the driving period. This approach is known as the *rotating wave approximation*. We get

$$\begin{pmatrix} \dot{a} \\ \dot{b} \end{pmatrix} = -i\frac{A\Delta}{4\hbar\Delta E}\begin{pmatrix} b e^{i\delta\omega t} \\ a e^{-i\delta\omega t} \end{pmatrix}. \tag{2.20}$$

Introducing the notation

$$\Omega_R^{(0)} = \frac{A\Delta}{2\hbar\Delta E}, \tag{2.21}$$

from Eq. (2.20) we obtain the following equation

$$\ddot{a} - i\delta\omega\dot{a} + \frac{\Omega_R^{(0)2}}{4}a = 0. \tag{2.22}$$

With the substitution $a = \exp(i\kappa t)$ we obtain the quadratic equation in κ, the solution of which gives us $\kappa_{1,2} = \frac{\delta\omega}{2} \pm \frac{\Omega_R}{2}$, where we defined

$$\Omega_R = \sqrt{\Omega_R^{(0)2} + \delta\omega^2}. \tag{2.23}$$

The value $\Omega_R^{(0)}$ is called the *Rabi frequency*, and the value Ω_R is known as the *generalized Rabi frequency*. Then for a we obtain

$$a_{1,2} = A_{1,2}e^{i\frac{\delta\omega\pm\Omega_R}{2}t}, \tag{2.24}$$

where $A_{1,2}$ are constants, which we will find from the normalization condition. From the first equation of the system (2.20) we have

$$b_{1,2} = -A_{1,2}\frac{\delta\omega \pm \Omega_R}{\Omega_R^{(0)}}e^{i\frac{-\delta\omega\pm\Omega_R}{2}t} \equiv B_{1,2}e^{i\frac{-\delta\omega\pm\Omega_R}{2}t}. \qquad (2.25)$$

From the normalization condition $|a_{1,2}|^2 + |b_{1,2}|^2 = 1$, taking into account (2.25), we find

$$A_{1,2} = \frac{1}{\sqrt{2}}\frac{\Omega_R^{(0)}}{\Omega_R}\left(1 \pm \frac{\delta\omega}{\Omega_R}\right)^{-1/2}, \quad B_{1,2} = \mp\frac{1}{\sqrt{2}}\left(1 \pm \frac{\delta\omega}{\Omega_R}\right)^{1/2}. \qquad (2.26)$$

So, from Eq. (2.18) we obtain the basis wave functions and the expansion in them:

$$|\psi'_{1,2}\rangle = \begin{pmatrix} a_{1,2}e^{i\frac{\omega_q}{2}t} \\ b_{1,2}e^{-i\frac{\omega_q}{2}t} \end{pmatrix} = e^{\pm i\frac{\Omega_R}{2}t}\begin{pmatrix} A_{1,2}e^{i\frac{\omega}{2}t} \\ B_{1,2}e^{-i\frac{\omega}{2}t} \end{pmatrix}, \qquad (2.27)$$

$$|\psi'\rangle = C_1|\psi'_1\rangle + C_2|\psi'_2\rangle.$$

Assume that in the initial moment of time, the system was in the ground state, $|\psi'\rangle = |E_-\rangle$. We find the expansion coefficients from the system of equations

$$\begin{pmatrix} 1 \\ 0 \end{pmatrix} = \begin{pmatrix} C_1A_1 + C_2A_2 \\ C_1B_1 + C_2B_2 \end{pmatrix}. \qquad (2.28)$$

Taking into account that $A_1B_2 - B_1A_2 = 1$ (which is easy to check), we obtain $C_1 = B_2$ and $C_2 = -B_1$. In order to find the probability of the system in the excited or ground state, we expand the wave function in them: $|\psi'\rangle = C_-|E_-\rangle + C_+|E_+\rangle = \begin{pmatrix} C_- \\ C_+ \end{pmatrix}$. Writing down the expression for C_+, which is the bottom element of the column (2.27), we obtain

$$P_+(t) = |C_+|^2 = \left| B_1 B_2 2i \sin \frac{\Omega_R}{2} t \right|^2 = \frac{\Omega_R^{(0)2}}{\Omega_R^2} \sin^2 \frac{\Omega_R}{2} t$$

$$= \frac{1}{2} \frac{\Omega_R^{(0)2}}{\Omega_R^2} (1 - \cos \Omega_R t). \tag{2.29}$$

This remarkable result shows the *Rabi oscillations* of the upper-level occupation probability. This means that the system experiences the oscillations with the period $T_R = 2\pi/\Omega_R$, when with the unitary probability it may appear in either the ground or excited state; see Fig. 2.2(a). At this, the oscillation frequency is proportional to the driving amplitude and is much smaller than the driving frequency:

$$\Omega_R \approx \Omega_R^{(0)} = \frac{A\Delta}{2\hbar\Delta E} \sim \frac{A}{\hbar} \ll \omega. \tag{2.30}$$

The time-averaged probability is described by the Lorentzian shape:

$$\overline{P_+} \equiv \frac{1}{T_R} \int_0^{T_R} dt P_+(t) = \frac{1}{2} \frac{\Omega_R^{(0)2}}{\Omega_R^{(0)2} + (\omega - \omega_q)^2}. \tag{2.31}$$

This means that at $\omega = \omega_q(\varepsilon_0^*)$ there is the resonance and the average excitation probability has the maximum, $\frac{1}{2}$, see Fig. 2.2(b). Note this fact. Namely, if we take into account that $P_- = 1 - P_+$, then we obtain that $\bar{P}_- \geq \bar{P}_+$, which means that it is impossible to create the inverse occupation in a periodically driven two-level system.

If we write the resonance condition in the form $\Delta E = \hbar\omega$, then the resonance can be interpreted as an exchange of one photon between the external field and our two-level system. The next subsection is devoted to the analogous multi-photon processes. Positions of one- and multi-photon resonances are shown in Fig. 2.2(c). Observation of such resonances in an experiment can be used for the definition of the distance between the levels. Such measurements are called *spectroscopy*: having the fixed frequency one can define the

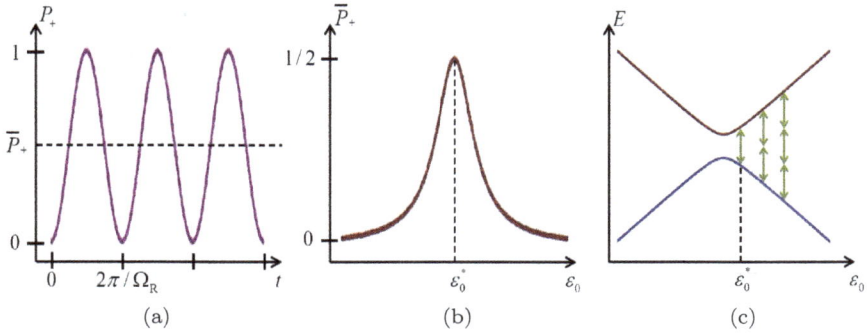

Fig. 2.2. Resonant excitation of a two-level system: (a) Rabi oscillations, (b) average upper-level excitation probability, (c) the energy levels and the positions of the one- and multi-photon resonances, defined by the relations $\Delta E(\varepsilon_0^*) = \hbar\omega$ and $\Delta E(\varepsilon_{0,k}^*) = k \cdot \hbar\omega$, respectively.

energy-level structure. Besides, from the half width at half maximum (HWHM), i.e. where $\overline{P_+} = \frac{1}{4}$, one can define the driving amplitude: $\delta\omega^{HWHM} = \Omega_R^{(0)} \propto A$.

2.3. Multi-photon excitations

2.3.1. *Schrödinger equation*

Consider now the strong excitation of a two-level system with the tunneling amplitude Δ a small value. That is when the following conditions are fulfilled

$$\Delta \ll \sqrt{A \cdot \hbar\omega}, \quad k\hbar\omega \sim \Delta E, \tag{2.32}$$

where the latter condition indicates the proximity of the energy of k photons to the qubit energy, $\hbar\omega_q = \Delta E \approx |\varepsilon_0|$. The small parameter of the problem is $\Delta^2/A\hbar\omega$; in the next Section we will see that this defines the adiabaticity parameter, and hence here we consider the case of the fast excitation. This means that the driving frequency is sufficiently large. The former condition can be considered as the condition on the driving amplitude (it is sufficiently large) or as the condition on the value Δ (its smallness allows the assumption $\Delta E \approx |\varepsilon_0|$).

Consider the Hamiltonian (2.3). Let us make the transformation to the rotating coordinate system by means of the operator

$$U(t) = \exp\left(-\frac{i}{\hbar}\int V(t)dt\right) \equiv \exp\left(i\frac{\eta(t)}{2}\sigma_z\right) = \cos\frac{\eta}{2} + i\sigma_z\sin\frac{\eta}{2},$$
(2.33)

$$\eta(t) = \frac{A}{\hbar\omega}\sin\omega t.$$

This operator links the wave functions ψ' in the rotating system and the ones in the original system $\psi : \psi = U(t)\psi'$. Then the Schrödinger equation $i\hbar\dot{\psi} = H\psi$ with the substitution $\psi = U(t)\psi'$ takes the form: $i\hbar U\dot{\psi}' + i\hbar\dot{U}\psi' = HU\psi'$. So, in the new system, we have $i\hbar\dot{\psi}' = H'\psi'$,

$$H' = U^\dagger H U - i\hbar U^\dagger\dot{U} = U^\dagger H U - U^\dagger V(t)U = U^\dagger H_0 U$$

$$= -\frac{1}{2}\begin{pmatrix} \varepsilon_0 & \Delta e^{-i\eta} \\ \Delta e^{i\eta} & -\varepsilon_0 \end{pmatrix} = -\frac{\Delta}{2}(e^{-i\eta}\sigma_+ + e^{i\eta}\sigma_-) - \frac{\varepsilon_0}{2}\sigma_z,$$
(2.34)

where $\sigma_\pm = \frac{1}{2}(\sigma_x \pm i\sigma_y)$, that is $\sigma_+ = \begin{pmatrix} 0 & 1 \\ 0 & 0 \end{pmatrix}$ and $\sigma_- = \begin{pmatrix} 0 & 0 \\ 1 & 0 \end{pmatrix}$.

We now use the *Jacobi–Anger expansion*,

$$e^{ix\sin\tau} = \sum_{n=-\infty}^{\infty} J_n(x)\, e^{in\tau},$$
(2.35)

where $J_n(x)$ is the first-kind Bessel function. Then the new Hamiltonian takes the form

$$H' = -\sum_{n=-\infty}^{\infty} \frac{\Delta_n}{2}(e^{-in\omega t}\sigma_+ + h.c.) - \frac{\varepsilon_0}{2}\sigma_z,$$
(2.36)

$$\Delta_n = \Delta J_n(A/\hbar\omega).$$

We note that, of course, the unitary transformation U does not change the level occupations (which are the absolute values of the spinor components):

$$\psi = \begin{pmatrix} \psi_1 \\ \psi_2 \end{pmatrix} = U\psi' = e^{i\frac{\eta}{2}\sigma_z}\begin{pmatrix} \psi_1' \\ \psi_2' \end{pmatrix} = \begin{pmatrix} e^{i\frac{\eta}{2}}\psi_1' \\ e^{-i\frac{\eta}{2}}\psi_2' \end{pmatrix}. \tag{2.37}$$

Here we first consider the evolution of a two-level system without relaxation. We look for a solution of the Schrödinger equation $i\hbar\dot{\psi}' = H'\psi'$ in the form

$$\psi' = \begin{pmatrix} \psi_1'' \exp\left(-i\dfrac{k\omega t}{2}\right) \\ \psi_2'' \exp\left(i\dfrac{k\omega t}{2}\right) \end{pmatrix}. \tag{2.38}$$

This corresponds to the transformation: $\psi'' = \exp(i\frac{k\omega t}{2}\sigma_z)\psi'$. We obtain the equations for ψ_1'' and ψ_2'', which are solved absolutely analogously to how we did this above. Consider the parameters in the vicinity of a k-photon resonance, where $k\hbar\omega \approx \Delta E \approx |\varepsilon_0|$. In other words, we now consider the case with small frequency detuning $\delta\omega_k = k\omega - |\varepsilon_0|/\hbar$. The rotating-wave approximation consists of the assumption that in the vicinity of a k-photon resonance, the fast-oscillating terms with $n \neq k$ can be neglected. Then it is straightforward to solve the Schrödinger equation analogously to how we did before for a one-photon excitation. As the result, we find the probability of the upper state, assuming that the system initially was in the ground state:

$$P_+^{(k)}(t) = |\psi_2(t)|^2 = |\psi_2''(t)|^2 = \overline{P_+^{(k)}}(1 - \cos\Omega_{R,k}t),$$

$$\tag{2.39}$$

$$\Omega_{R,k} = \frac{1}{\hbar}\sqrt{\Delta_k^2 + (k\hbar\omega - |\varepsilon_0|)^2}, \quad \overline{P_+^{(k)}} = \frac{1}{2}\frac{\Delta_k^2}{\Delta_k^2 + (k\hbar\omega - |\varepsilon_0|)^2}.$$

For $A \ll \hbar\omega$ we can use the asymptote of the Bessel function: $J_k(x) \approx \frac{x^k}{2^k k!}$. Remarkably, at small driving amplitudes A, the HWHM is proportional to A^k, while for increased driving amplitudes the HWHM is on the order of Δ. Furthermore, at the point of resonance we have $\hbar\Omega_{R,1} = \Delta_1 \approx \Delta A/2\hbar\omega$; and this expression coincides with the one obtained above for the Rabi frequency, Eq. (2.21), if we take into account that in the resonance $\hbar\omega = |\varepsilon_0| \approx \Delta E$. This is a remarkable and pedagogic result: look, we have considered here the regime of the strong excitation that is of high driving amplitude A, and then, formally, we have shifted to the limit of the small amplitude. And with this we have obtained the correct answer, coinciding with the result of the previous subsection!

2.3.2. *Liouville–von Neumann equation*

Let us now take the relaxation into account. We will describe the system by the Bloch equations, which include two phenomenological relaxation time parameters, T_1 and T_2. For the description of the dissipative dynamics of a quantum system, it is convenient to use the density matrix formalism [Blum 1981]. So, we will introduce here the density matrix and consider the equation for it, first without dissipation.

It was assumed above that a quantum system is described by a wave function $|\psi\rangle$ — such states are called *pure states*. But often we have the situation where a system's wave function is not known and we know only that the system with the set of probabilities p_i is in the states of the statistical ensemble of pure states $|\psi_i\rangle$. Such a state is called a *statistical mixture* or a *mixed state*. Accordingly, the density operator for a pure state is defined as $\rho = |\psi\rangle\langle\psi|$, and for a mixed state: $\rho = \sum p_i|\psi_i\rangle\langle\psi_i|$.

From the definition of a density operator, the following properties follow. It is Hermitian, $\rho^\dagger = \rho$. With this density operator, the system can be fully described, since it defines the observable values, $\langle A \rangle = Sp(\rho A)$; at this $Sp(\rho) = 1$ and $Sp(\rho^2) \leq 1$, where the equality sign corresponds to a pure state, for which $\rho^2 = \rho$. For the description of the density matrix, consider this in a specific basis $\{|e_n\rangle\}$:

$$|\psi_i\rangle = \sum_n a_n^{(i)}|e_n\rangle,$$

$$\rho_{nm} = \langle e_n|\rho|e_m\rangle = \sum_i p_i\langle e_n|\psi_i\rangle\langle\psi_i|e_m\rangle = \sum_i p_i a_n^{(i)} a_m^{(i)*}. \tag{2.40}$$

A diagonal element of the density matrix $\rho_{nn} = \sum_i p_i |a_n^{(i)}|^2$ equals the probability of the state $|e_n\rangle$ in the i-th component of the statistical ensemble, averaged over the ensemble. So, the diagonal components of the density matrix ρ_{nn} give the probability of observing the system in the state $|e_n\rangle$. They are called *occupations* of the respective states. Off-diagonal components contain the cross terms $a_n^{(i)} a_m^{(i)*}$ and reflect interference phenomena between the states $|e_n\rangle$ and $|e_m\rangle$ in the coherent superposition of these states in the i-th component of the statistical ensemble. For this reason, the off-diagonal components of a density matrix are called *coherences*.

The equation for the density operator has to be fulfilled both for the mixed and the pure states. For this reason we can allow ourselves to consider the simpler case of a pure state:

$$\frac{d}{dt}\rho = \frac{d}{dt}|\psi\rangle\langle\psi| = \frac{1}{i\hbar}H|\psi\rangle\langle\psi| - \frac{1}{i\hbar}|\psi\rangle\langle\psi|H = -\frac{i}{\hbar}[H,\rho]. \quad (2.41)$$

This equation is called the *Liouville–von Neumann equation*. (This equation is also called either quantum Liouville equation or von Neumann equation.)

Consider the Liouville–von Neumann equation for the case of a single qubit. It is convenient to use the parametrization for the density matrix in the form of an expansion in the Pauli matrices:

$$\rho = \frac{1}{2}(1\sigma_0 + X\sigma_x + Y\sigma_y + Z\sigma_z) = \frac{1}{2}\begin{pmatrix} 1+Z & X-iY \\ X+iY & 1-Z \end{pmatrix}$$

$$\equiv \begin{pmatrix} \rho_{00} & \rho_{01} \\ \rho_{10} & \rho_{11} \end{pmatrix}. \quad (2.42)$$

From the hermiticity requirement it follows that $\rho_{01} = \rho_{10}^*$, and from the normalization requirement it follows that $\rho_{00} + \rho_{11} = 1$.

It is useful to rewrite the Liouville–von Neumann equation for a single qubit in the form of the so-called Bloch equation. For this, we define the *Bloch vector*: $\vec{R} = (X, Y, Z)$. Then for the qubit density matrix we have: $\rho = \frac{1}{2}(\sigma_0 + \vec{R}\vec{\sigma})$. Let us write down the Hamiltonian analogously, $H = -\frac{\hbar}{2}\vec{H}\vec{\sigma}$, where we have defined the effective "magnetic field", in analogy with the Hamiltonian of the spin $1/2$ in a magnetic field, $H = -\vec{\mu}\vec{H}$, see Eq. (1.6). In the case of the qubit Hamiltonian (2.1) we have $\vec{H} = (\Delta/\hbar, 0, \varepsilon/\hbar)$. Then for the density matrix in a pure state we have:

$$|\psi\rangle = \begin{pmatrix} \cos\frac{\theta}{2} \\ e^{i\phi}\sin\frac{\theta}{2} \end{pmatrix} \quad \Rightarrow \quad \rho = |\psi\rangle\langle\psi| = \frac{1}{2}\begin{pmatrix} 1 + \cos\theta & e^{-i\phi}\sin\theta \\ e^{i\phi}\sin\theta & 1 - \cos\theta \end{pmatrix}.$$

(2.43)

This means that the components of the Bloch vector are defined by the polar and azimuthal angles:

$$\vec{R} = (\sin\theta\cos\phi, \sin\theta\sin\phi, \cos\theta). \tag{2.44}$$

Then the Liouville–von Neumann equation takes the form

$$\frac{1}{2}\dot{\vec{R}}\vec{\sigma} = -\frac{i}{\hbar}(H\rho - \rho H) = \frac{i}{4}[(\vec{H}\vec{\sigma})(\vec{R}\vec{\sigma}) - (\vec{R}\vec{\sigma})(\vec{H}\vec{\sigma})]$$

$$= \frac{i}{4}[H_i R_j \sigma_i \sigma_j - H_i R_j \sigma_j \sigma_i] = \frac{i}{4}H_i R_j [\sigma_i, \sigma_j]. \tag{2.45}$$

In view that

$$\sigma_i \sigma_j = \delta_{ij} + i\varepsilon_{ijk}\sigma_k, \quad [\sigma_i, \sigma_j] = i2\varepsilon_{ijk}\sigma_k, \tag{2.46}$$

we obtain

$$\dot{\vec{R}}\vec{\sigma} = -\varepsilon_{ijk}H_i\rho_j\sigma_k = -\vec{H} \times \vec{R} \cdot \vec{\sigma}. \tag{2.47}$$

From this relation, we have the equation for the vector $\vec{R}$:

$$\dot{\vec{R}} = \vec{R} \times \vec{H}. \tag{2.48}$$

This equation formally coincides with the Bloch equations, without relaxation for the evolution of a magnetic moment in a magnetic field. This evolution corresponds to the Larmor precession of a vector $\vec{R}$ around the effective magnetic field $\vec{H}$ with the angular frequency $|\vec{H}| = \frac{1}{\hbar}\sqrt{\Delta^2 + \varepsilon_0^2} = \omega_q$. (In order to see this, one can point the z axis along $\vec{H}$, then it follows that $R_z = const$ and $R_x \propto \sin\omega_q t$, $R_y \propto \cos\omega_q t$.) So, in the absence of relaxation, the dynamics of a system's pure state is described by the precession of the vector $\vec{R}$ around the vector $\vec{H}$ on the unitary Bloch sphere. In particular, if the vector $\vec{H}$

coincides with the x axis, i.e. $\varepsilon_0 = 0$, then the precession takes place in the (y, z) plane and the system periodically passes the north and south poles, which means that it oscillates between the states $|0\rangle$ and $|1\rangle$ — these are the so-called *quantum beats*. So, the quantum beats allow the transition between the basis (physical) states to be realized. However, if the qubit is initially in equilibrium, with $\vec{R} \parallel \vec{H}$, then it would stay therein. In this case, the transition between the basis states can be realized by applying the external periodic field, then we obtain the Rabi oscillations, or by rapidly changing the Hamiltonian parameters, i.e. by applying rectangular pulses.[16]

Consider once again the qubit evolution in the absence of any external perturbation — now by solving the Liouville–von Neumann equation in the energy representation with the Hamiltonian $H_0 = -\frac{\Delta E}{2}\sigma_z = E_-|E_-\rangle\langle E_-| + E_+|E_+\rangle\langle E_+|$, where $E_\pm = \pm\Delta E/2$. In this representation, from the Liouville–von Neumann equation, we obtain:

$$\begin{cases} \dot{\rho}_{00} = 0, \\ \dot{\rho}_{10} = -\dfrac{i}{\hbar}\Delta E\rho_{10}. \end{cases} \tag{2.49}$$

The solution of these equations is $\rho_{00} = const$ and $\rho_{10} = \rho_{10}(0)\exp\left(-\frac{i}{\hbar}\Delta Et\right)$. These relations, once again, describe the quantum beats.

2.3.3. *Bloch equations*

If we need to describe the impact of the environment, i.e. if our system interacts with the reservoir, then the general approach consists of the consideration of the Liouville–von Neumann equation for the aggregate system. This equation is then averaged over the reservoir degrees of freedom. After certain transformations and simplifications, the Master kinetic equation is obtained. Section 2.5 is devoted to this approach. Consider here, first, the relaxation phenomena rather phenomenologically. For this, in order to obtain the decaying

---

[16]G. Wendin and V. S. Shumeiko, Superconducting quantum circuits, qubits and computing, arXiv:cond-mat/0508729 (2005).

solution, we have to add, in the r.h.s. of the equation $\dot{\rho}_{ij} = -\frac{i}{\hbar}[H, \rho]_{ij}$, the additional term, $-\frac{\rho_{ij} - \rho_{ij}^{(0)}}{\tau_{ij}}$, and then we have

$$\dot{\rho}_{ij} = -\frac{i}{\hbar}[H, \rho]_{ij} - \frac{\rho_{ij} - \rho_{ij}^{(0)}}{\tau_{ij}}. \tag{2.50}$$

Indeed, the solution of the equation $\dot{\rho}_{ij} = -\frac{\rho_{ij} - \rho_{ij}^{(0)}}{\tau_{ij}}$ is $\rho_{ij}(t) = \rho_{ij}^{(0)} + Ce^{-t/\tau_{ij}}$. Let us define further the equilibrium values for the density matrix.

Given that our system is at thermal equilibrium with the environment at temperature T, as it is known from the course on statistical physics, the equilibrium density operator is $\rho^{(0)} = \exp(-H_0/k_B T)/\Sigma$, where $k_B = 1.38 \cdot 10^{-23}$ J/K is the Boltzmann constant. Here, the statistical sum $\Sigma = Tr\exp(-H_0/k_B T)$ plays the role of the normalizing factor. Then for the matrix elements we have

$$\rho_{00/11}^{(0)} = \frac{1}{\Sigma}\langle E_\mp | e^{-H_0/k_B T} | E_\mp \rangle = \frac{1}{\Sigma}e^{-E_\mp/k_B T}$$

$$= \frac{1}{\Sigma}e^{\pm \Delta E/2k_B T}, \quad \rho_{10}^{(0)} = 0. \tag{2.51}$$

This means that the populations are defined by the Boltzmann distribution. We find the normalizing factor from the condition $1 = \rho_{00}^{(0)} + \rho_{11}^{(0)} = \frac{2}{\Sigma}\cosh\frac{\Delta E}{2k_B T}$. From this, we obtain the value for the parameter Z of the density matrix in equilibrium, which defines the difference in the level occupations:

$$Z^{(0)} = \rho_{00}^{(0)} - \rho_{11}^{(0)} = \tanh\frac{\Delta E}{2k_B T}. \tag{2.52}$$

In particular, at low temperatures, we have $Z^{(0)} \approx 1$.

So, looking at Eq. (2.50), in the r.h.s. of the equation for Z we have to add $(Z - Z^{(0)})/T_1$ and for the off-diagonal terms (that is for X and Y) we have to add a term of the form ρ_{10}/T_2. Here T_1 and T_2 are phenomenological *energy relaxation time* and *decoherence time*, respectively. The Liouville–von Neumann equation with such additions results in the so-called *Bloch equations*. Let us write them down here separately for diagonal and off-diagonal density-matrix components, respectively:

$$\dot{\rho}_{ii} = -\frac{i}{\hbar}[H, \rho]_{ii} - \frac{\rho_{ii} - \rho_{ii}^{(0)}}{T_1},$$

$$\dot{\rho}_{ij} = -\frac{i}{\hbar}[H, \rho]_{ij} - \frac{\rho_{ij}}{T_2}. \tag{2.53}$$

We note that for the off-diagonal terms we have $\rho_{ji} = \rho_{ij}^*$, and that, instead of the diagonal terms, it is sometimes more convenient to use their difference $Z = \rho_{00} - \rho_{11}$.

In our case, we have to put the Hamiltonian (2.36) in the equation. Leaving only the states with $n = k$ in the rotating-wave approximation (as it was discussed above), we write down the Hamiltonian (2.36) in the form

$$H' = -\frac{1}{2}\begin{pmatrix} \varepsilon_0 & \Delta_k e^{-ik\omega t} \\ \Delta_k e^{ik\omega t} & -\varepsilon_0 \end{pmatrix}, \quad \Delta_k = \Delta J_k\left(\frac{A}{\hbar\omega}\right). \tag{2.54}$$

Then the Bloch equations (2.53) take the form:

$$\dot{\rho}_{10} = i\frac{\Delta_k}{2\hbar}e^{ik\omega t}Z - i\frac{\varepsilon_0}{\hbar}\rho_{10} - \frac{\rho_{10}}{T_2},$$

$$\dot{Z} = i\frac{\Delta_k}{\hbar}(e^{-ik\omega t}\rho_{10} - \text{c.c.}) - \frac{Z - Z^{(0)}}{T_1}. \tag{2.55}$$

After the substitution $\rho_{10}\exp(-ik\omega t) = \frac{1}{2}(\tilde{X} + i\tilde{Y})$, we obtain the following system of equations

$$\dot{\tilde{X}} = \left(k\omega + \frac{\varepsilon_0}{\hbar}\right)\tilde{Y} - \frac{\tilde{X}}{T_2},$$

$$\dot{\tilde{Y}} = -\left(k\omega + \frac{\varepsilon_0}{\hbar}\right)\tilde{X} + \frac{\Delta_k}{\hbar}Z - \frac{\tilde{Y}}{T_2}, \tag{2.56}$$

$$\dot{Z} = -\frac{\Delta_k}{\hbar}\tilde{Y} - \frac{Z - Z^{(0)}}{T_1}.$$

In equilibrium, the solution of these equations corresponds to equating the time derivatives in the l.h.s. of Eq. (2.56) to zero. Then for the stationary value of the upper-level occupation, $\overline{P}_+^{(k)} = \overline{\rho}_{11}^{(k)} = \frac{1}{2}(1 - \overline{Z}^{(k)})$, we obtain at low temperature (when $Z^{(0)} \approx 1$, see Eq. (2.52)) the following formula

$$\overline{P}_+(\varepsilon_0, A) = \sum_k \overline{P}_+^{(k)} = \frac{1}{2} \sum_k \frac{\Delta_k^2}{\Delta_k^2 + \frac{T_2}{T_1}(k\hbar\omega - \varepsilon_0)^2 + \frac{\hbar^2}{T_1 T_2}},$$

$$(2.57)$$

where we have replaced the summation index $k \to -k$. This formula is useful for the description of the multi-photon resonances in a two-level system. Note that at $T_1 = T_2 \to \infty$ this formula coincides with the one obtained above, Eq. (2.39). We can see from Eq. (2.57) that the upper-level occupation is maximal at $\varepsilon_0 = k \cdot \hbar\omega$, which corresponds to the k-photon resonant transitions between the energy levels, see Fig. 2.3(a). The width of the resonance lines is defined by the numerator, which contains the Bessel function. Note that the asymptotics of the Bessel function, at large values of its argument, have the form: $J_k(x) \approx \sqrt{\frac{2}{\pi x}} \cos\left(x - \frac{\pi}{4}(2k + 1)\right)$; this is the function, of which the second power (cf. Δ_k^2) is quasi-periodic with the period π. The positions of the Bessel-function zeros are shown by the crosses in Fig. 2.3(a).

In Fig. 2.3(b) we present the *interferogram* for $\overline{P}_+$ in the strong-driving regime, calculated with Eq. (2.57) for $\Delta/\hbar\omega = 0.8$. Such interferograms are useful as tools for studying realistic systems: (a) one can define the system spectrum from the position of resonances (multi-photon spectroscopy), (b) the driving amplitude can be defined from the distance between the Bessel-function zeros (power calibration), and (c) analyzing the shape of the resonances, one can obtain the temperature and the relaxation times, parameters which relate to the environment. Such an approach to the description of a quantum system is known as *Landau–Zener–Stückelberg–Majorana interferometry*.[17] This name came about because the resonant excitation of a system can be explained as a result of the Landau–Zener transitions (a.k.a. Landau–Zener–Stueckelberg–Majorana transitions) between the system energy levels together

[17]S. N. Shevchenko, S. Ashhab, and F. Nori, Landau–Zener–Stückelberg interferometry, Phys. Reports **492**, 1 (2010).

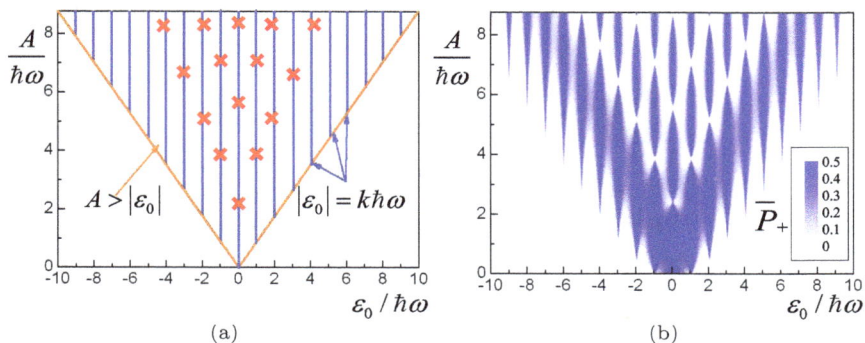

Fig. 2.3.   Illustration of the Landau–Zener–Stückelberg–Majorana interferometry: (a) position of the resonances and (b) dependence of the time-averaged upper-level occupation probability $\overline{P}_+$ on ε_0 and A. The restriction $A > |\varepsilon_0|$ will be explained later. The data in panel (b) are calculated with the formula (2.57). We can see several multi-photon resonances at $\varepsilon_0 = k\hbar\omega$ for $k = 0, 1, \cdots 9$. Their width is modulated by the Bessel functions. The important role is played by the distance between the successive zeros of a Bessel function: they are marked by crosses in (a) and they are visible as interruptions in the resonance lines in (b).

with the accumulation of the Stückelberg phase. These fundamental phenomena are related to various physical systems; therefore the next section, §2.4, is devoted to their detailed consideration.

2.4.   Landau–Zener transitions and Stückelberg oscillations

So, consider here one more approach to the solution of the problem about the evolution of a two-level system (qubit), based on the notion of the adiabatic evolution. This means that we will discuss quite slow driving, with small frequency. Incidentally, the obtained results describe correctly the problem with fast evolution as well (See Footnote 17 on page 54).

2.4.1.   *Adiabatic energy levels*

Consider the adiabatic evolution when a system evolves in one of its adiabatic states, which are defined as the eigenstates of $H(t)$. This means that the *adiabatic basis* consists of the instantaneous eigenstates of the time-dependent Hamiltonian. This corresponds to changing the problem on the eigenstates and eigenvalues in §2.1, formulated as $H_0|E_\pm\rangle = E_\pm|E_\pm\rangle$, to the following

$$H(t)|E_\pm(t)\rangle = E_\pm(t)|E_\pm(t)\rangle. \qquad (2.58)$$

Here, we may not solve this problem repeatedly, but rather use the results from §2.1 with the substitution $\varepsilon_0 \to \varepsilon(t)$. In particular, for the adiabatic energy levels we obtain

$$E_\pm(t) = \pm\frac{1}{2}\sqrt{\Delta^2 + \varepsilon(t)^2} \equiv \pm\frac{1}{2}\Delta E(t). \qquad (2.59)$$

The non-stationary Schrödinger equation $i\hbar\frac{\partial}{\partial t}|E_\pm(t)\rangle = H(t)|E_\pm(t)\rangle$, together with Eq. (2.58), give

$$|E_\pm(t)\rangle = \exp\left(-\frac{i}{\hbar}\int_0^t E_\pm(t')dt'\right)|E_\pm(0)\rangle = e^{\mp i\zeta(t)}|E_\pm(0)\rangle,$$

$$\zeta(t) = \frac{1}{2\hbar}\int_0^t \Delta E(t')dt'. \qquad (2.60)$$

{The account of the next-order terms in the quasi-classical (adiabatic) approximation results in an additional $\pi/4$ to the phase (See Footnote 17 on page 54).} Then, a wave function, which describes the quantum state of a two-level system as a function of time, can be expanded in the adiabatic basis,

$$|\Psi(t)\rangle = \sum_\pm c_\pm(t)e^{\mp i\zeta(t)}|E_\pm(0)\rangle \equiv \sum_\pm b_\pm(t)|E_\pm(0)\rangle. \qquad (2.61)$$

Consider adiabatic evolution. Since in the adiabatic approximation the coefficients $c_\pm$ in Eq. (2.61) do not depend on time between the quasi-crossings (i.e. the system does not pass from one adiabatic state to another), the adiabatic evolution from $t = t_i$ to $t = t_f$ can be described as the following: the coefficients $b_\pm(t_i) = c_\pm e^{\mp i\zeta(t_i)}$ become $b_\pm(t_f) = c_\pm e^{\mp i\zeta(t_f)}$. Here we can rewrite the phase, $\zeta(t_f) = \zeta(t_f) - \zeta(t_i) + \zeta(t_i) \equiv \Delta\zeta + \zeta(t_i)$, then $b_\pm(t_f) = e^{\mp i\Delta\zeta}b_\pm(t_i)$ (note that $|b_\pm(t)|$ do not change). This can be conveniently written in the matrix form, after introducing the matrix U for the adiabatic evolution:

$$\mathbf{b}(t_\mathrm{f}) = U(t_\mathrm{f}, t_\mathrm{i})\mathbf{b}(t_\mathrm{i}),$$

$$\mathbf{b}(t) \equiv \begin{pmatrix} b_-(t) \\ b_+(t) \end{pmatrix},$$

$$U(t_\mathrm{f}, t_\mathrm{i}) = \begin{pmatrix} e^{i\Delta\zeta} & 0 \\ 0 & e^{-i\Delta\zeta} \end{pmatrix} = e^{i\Delta\zeta\sigma_z}, \qquad (2.62)$$

$$\Delta\zeta = \frac{1}{2\hbar} \int_{t_\mathrm{i}}^{t_\mathrm{f}} \Delta E(t)\,dt.$$

Such a description, by means of the matrices linking the wave-function amplitudes, is called the *transfer-matrix method*.

Let us describe the evolution graphically. The ε-dependent energy levels are shown in Fig. 2.4. The important point is the approach of the levels at $\varepsilon = 0$, which is also called *quasi-crossing* or *avoided-level crossing*. Two solid lines represent the adiabatic energy levels.

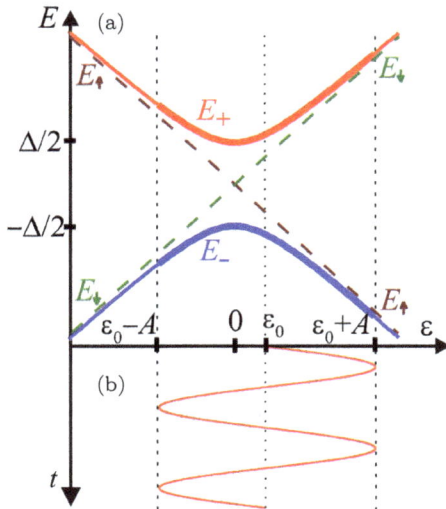

Fig. 2.4.   (a) The energy levels E in dependence on the bias ε. (b) The bias ε represents the driving signal $\varepsilon(t) = \varepsilon_0 + A \sin \omega t$, which oscillates between $\varepsilon_\mathrm{min} = \varepsilon_0 - A$ and $\varepsilon_\mathrm{max} = \varepsilon_0 + A$.

Dashed lines are the diabatic energy levels $E_{\uparrow,\downarrow}$, corresponding to the diabatic states. The *diabatic* basis is formed by the eigenstates of the operator σ_z. Note that these states become the eigenstates of the Hamiltonian, if Δ is neglected.

Next, for simplification, consider the case of the zero offset $\varepsilon_0 = 0$. The adiabatic energy levels $E_\pm(t)$ have minimal distance Δ, which is realized in the instants of time $\pi n/\omega$, where n is an integer; see Fig. 2.5. Since the adiabatic states quickly change with time in the region of the quasi-crossing and slowly change out of this region, we can describe the evolution in this latter case adiabatically, with the probability of non-adiabatic transitions in the vicinity of the quasi-crossing points. It is convenient not to detach this region of

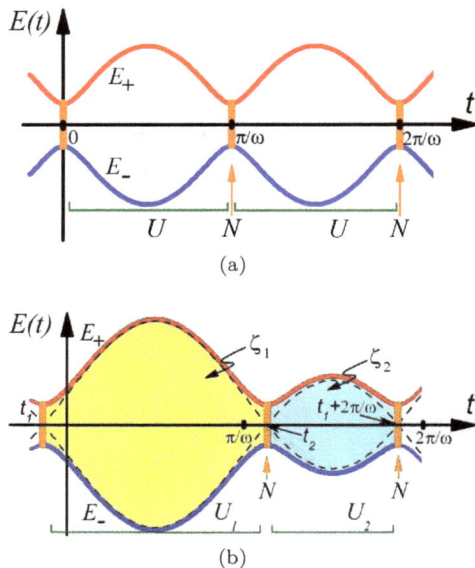

Fig. 2.5.   Time dependence of the energy levels during one period. (a) The case of zero offset, $\varepsilon_0 = 0$, which is considered in the text, and (b) more general case of nonzero offset, $\varepsilon_0 \neq 0$, studied in Footnote 17 (page 54). The time-dependent energy levels define the two-stage evolution: transitions in the non-adiabatic regions are defined by the evolution matrix N, while the adiabatic evolution is defined by the matrices $U_{1,2} = \exp(-i\zeta_{1,2}\sigma_z)$. The phase changes $\zeta_{1,2}$ have geometric interpretation: they are equal to the areas under the curves during the respective time span. The diabatic energy levels, $\pm\varepsilon(t)/2$, are shown by the dashed lines.

non-adiabaticity, but rather to attribute the transition probability to this only point, where the levels maximally approach each other. Such a discretized (stroboscopic) picture significantly simplifies the calculations; this was strictly grounded in Footnote 17 (page 54). We call such description the *adiabatic-impulse approximation*, so as to emphasize the two-stage character of the evolution. In particular, this term emphasizes that the non-adiabatic transitions are described as instantaneous ones, that is, in essence, only a convenient trick for describing the dynamics.

Within the framework of the adiabatic-impulse approximation, we can understand the condition for the appearance of resonances, at $|\varepsilon_0| < A$, which was used for plotting Fig. 2.3(a). As one can see from Fig. 2.4, the point at which levels approaching is crossed, only if $|\varepsilon_0| < A$. Therefore, only in this case, we can expect transitions to the upper level. This is remarkably confirmed by the numerical solution of the Schrödinger equation, presented in Fig. 2.3(b).

2.4.2. *Single passage: Landau–Zener transition*

Consider a non-adiabatic region in the vicinity of the points of quasi-crossing: $t = \pi n/\omega + t', \omega|t'| \ll 1$. Then the bias can be linearized:

$$\varepsilon(\pi n/\omega + t') = \pm A \sin \omega t' \approx \pm v t', \quad v = A\omega. \tag{2.63}$$

The linear Hamiltonian $H(t) = -\frac{\Delta}{2}\sigma_x \mp \frac{vt}{2}\sigma_z$ represents the Landau–Zener problem. Here and below we omit the primes.

Consider the solution, which is based on the fact that the transitions with adiabatic excitation bear quasi-classical character, when the change of the action (given by the integral $\int E(t)dt$) is large [Landau and Lifshitz 1977, §53]. Then the problem regarding the transition with adiabatic perturbation is formally analogous to the problem of quasi-classical over-barrier reflection. Information about the other approaches to the solution of this problem can be found in many papers, see for example the reference in Footnote 18.

---

[18]F. Di Giacomo and E. E. Nikitin, The Majorana formula and the Landau–Zener–Stückelberg treatment of the avoided crossing problem, Phys. Uspekhi **48**, 515 (2005); F. Wilczek, Majorana and Condensed Matter Physics, arXiv:1404.0637 (2014).

Assuming that the system initially, at $t = -\infty$, was at the lower energy level, the probability of transition from the lower level to the upper one during the single passage is described by the quasi-classical approximation, as follows. The energy levels $E_\pm(t) = \pm\frac{1}{2}\sqrt{\Delta^2 + (vt)^2}$ coincide at two points of the complex plane: at $t = \pm i\Delta/v \equiv \pm t_0$; then $E_+(t_0) = E_-(t_0)$. Thereafter, the probability P_+ that the system appears in the upper level (at $t = +\infty$) is defined by the integral [Landau and Lifshitz 1977, §53]:

$$P_+ = \exp\left(-\frac{2}{\hbar}\text{Im}\int_0^{t_0} [E_+(t) - E_-(t)]dt\right). \qquad (2.64)$$

We have

$$\int_0^{t_0} \Delta E(t)dt = \frac{\Delta^2}{v}\int_0^i \sqrt{1 + z^2}dz = \frac{\Delta^2}{2v}\left(z\sqrt{1 + z^2} + \text{asinh}(z)\right)_0^i$$

$$= i\frac{\pi\Delta^2}{4v}, \qquad (2.65)$$

where we have calculated the integral

$$I = \int \sqrt{1 + z^2}dz = z\sqrt{1 + z^2} - \int \frac{z^2 + 1 - 1}{\sqrt{1 + z^2}}dz$$

$$= z\sqrt{1 + z^2} - I + \text{asinh}(z) \qquad (2.66)$$

and have taken into account that $\text{asinh}(i) = i\pi/2$. We obtain the Landau–Zener probability

$$P_+ = P_{\text{LZ}} = \exp(-2\pi\delta),$$

$$\delta = \frac{\Delta^2}{4\hbar v}. \qquad (2.67)$$

(We note that, while the equation (2.67) is most commonly attributed to being the Landau–Zener formula, this should be called the Landau–Zener–Stückelberg–Majorana formula (See Footnote 18 on page 59).) It happens that this formula describes the transition probability for arbitrary values of the exponent.[19] With a change in the excitation velocity v from 0 (the adiabatic limit) to ∞

---

[19]E. E. Nikitin and S. Ya. Umansky, Nonadiabatic transitions under slow atomic collisions (Atomizdat, Moscow, 1979).

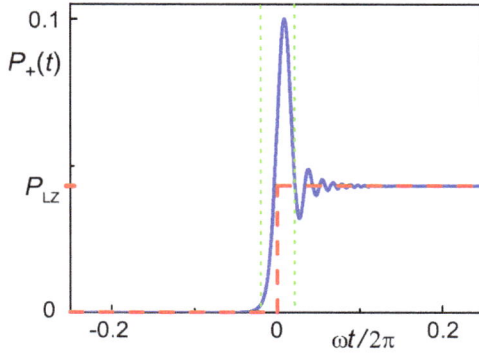

Fig. 2.6. Landau–Zener transition at the single passage of the "avoided-level-crossing point": the time dependence of the upper-level occupation probability $P_+(t)$. The solid line represents the numerical solution of the Schrödinger equation and the dashed line corresponds to the analytic solution.

(fast-changes limit), the transition probability P_+ varies from 0 to 1. The parameter δ is called the *adiabaticity parameter*. The statement that in the adiabatic limit, the transition probability equals to zero, is known as the *adiabaticity theorem*. In Fig. 2.6 the dashed line shows the instantaneous transition within the adiabatic-impulse model with the probability given by the formula (2.67). The solid line is obtained by numerical solution of the Schrödinger equation. Note the remarkable agreement.

The Landau–Zener formula (2.67) describes the occupation probability of the upper level P_+, which is the squared absolute value of the probability amplitude. However, when the interference is important, it is necessary to know also the changes of the respective phase of the two components of the spinor wave function in the process of passing the quasi-crossing region. Overall, the evolution process, at passing this region, is described by the unitary matrix, which for simplicity we write down here for the case of the adiabatic passage, at $\delta \gg 1$ (Footnote 17 on page 54):

$$\mathbf{b}(\pi n/\omega + 0) = N\mathbf{b}(\pi n/\omega - 0),$$

$$N = \begin{pmatrix} -ir & -t \\ t & ir \end{pmatrix}, \quad r = \sqrt{1 - P_{\mathrm{LZ}}}, \quad t = \sqrt{P_{\mathrm{LZ}}}. \tag{2.68}$$

Indeed, let initially the system be, before the transition, at the instant of time $t_i = \pi n/\omega - 0$ in the ground state,

$$\mathbf{b}(t_i) = \begin{pmatrix} b_-(t_i) \\ b_+(t_i) \end{pmatrix} = \begin{pmatrix} 1 \\ 0 \end{pmatrix}. \qquad (2.69)$$

Then, at $t_f = \pi n/\omega + 0$, from Eq. (2.68) we have the respective probabilities — to pass to the excited state or to stay in the ground state:

$$|b_+(t_f)|^2 = |t|^2 = P_{\mathrm{LZ}}, \quad |b_-(t_f)|^2 = |r|^2 = 1 - P_{\mathrm{LZ}}. \qquad (2.70)$$

So, we confirmed that the matrix N is defined so that the diagonal elements correspond (up to the phase factor) to the "reflection" and the off-diagonal elements correspond to the "transmission".

* In general case, for the matrix N we have

$$N = \begin{pmatrix} re^{i\varphi} & -t \\ t & re^{-i\varphi} \end{pmatrix},$$

$$\varphi = \varphi(\delta) = -\frac{\pi}{4} + \delta(\ln \delta - 1) + \arg \Gamma(1 - i\delta), \qquad (2.71)$$

where $\varphi(\delta)$ is the so-called *Stokes phase*, dependent on the adiabaticity parameter δ, and Γ is the gamma-function. The argument of the gamma-function asymptotics has the form

$$\arg\Gamma(1 - i\delta) \approx \begin{cases} C\delta, & \delta \ll 1, \\ -\pi/4 - \delta(\ln \delta - 1), & \delta \gg 1, \end{cases}$$

where $C \approx 0.58$ is the Euler constant. So, the jump of the phase during the transition $\varphi(\delta)$ is the monotonous function, changing between $-\pi/2$ in the adiabatic (slow) limit ($\delta \gg 1$) and $-\pi/4$ in the diabatic (fast) limit ($\delta \ll 1$). In the case considered above, we had $\delta \gg 1$ and $\varphi = -\pi/2$.

2.4.3. *Double passage: Stückelberg phase*

Consider now the process of repeated passage of the level quasi-crossing point. Such a problem corresponds to one full period of the periodic driving. Then, the adiabatic evolution is described by the matrix

$$U = e^{i\zeta\sigma_z}, \quad \zeta = \frac{1}{2\hbar} \int_0^{\pi/\omega} \Delta E(t)dt \tag{2.72}$$

and the non-adiabatic transitions are described by the matrix N.

Consider the total evolution matrix during one period — see Fig. 2.5 — this is obtained by simple multiplication of the four matrices:

$$NUNU = \begin{pmatrix} \beta & -\xi^* \\ \xi & \beta^* \end{pmatrix}, \quad \beta = -r^2 e^{i2\zeta} - t^2, \quad \xi = irt\left(e^{-i2\zeta} - 1\right). \tag{2.73}$$

This means that the evolution is described by the formula $\mathbf{b}(t_f) = NUNU\mathbf{b}(t_i)$. If, as above, the system was initially in the ground state, Eq. (2.69), then the excitation (transition) probability is defined by the off-diagonal matrix element, which equals to ξ:

$$P_2 \equiv |b_+(t_f)|^2 = |\xi|^2 = |rt(e^{i\zeta} - e^{-i\zeta})|^2 = 4P_{\mathrm{LZ}}(1 - P_{\mathrm{LZ}})\sin^2\zeta. \tag{2.74}$$

This result shows that the excitation probability is the oscillating function of the phase ζ. This was first studied by E. Stückelberg. In the majority of cases, the *Stückelberg oscillations* used to be unobservable, since the observable values were described by the average probability

$$\overline{P}_+ = 2P_{\mathrm{LZ}}(1 - P_{\mathrm{LZ}}); \tag{2.75}$$

see e.g. [Landau and Lifshitz 1977, §90]. This expression, as a matter of fact, is given by the sum of the two probabilities, which correspond to the possibilities of excitation during either the first or the second passage of the quasi-crossing point. These two possibilities are shown in Fig. 2.7 by the trajectories marked by single or double arrows, respectively.

The quantum-mechanical interference between different Landau–Zener transitions gives the result that the excitation probability after the repeated passage can have the value between 0 (destructive

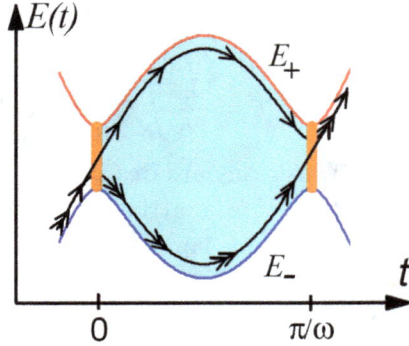

Fig. 2.7.   Double passage. The adiabatic energy levels are plotted as in Fig. 2.5; lines with single or double arrows show the trajectories, where the transition to the upper level happens during either the first or the second passage of the quasi-crossing points. The respective excitation probabilities are defined by the expressions $P_{LZ} \times (1 - P_{LZ})$ and $(1 - P_{LZ}) \times P_{LZ}$.

interference) and $4P_{LZ}(1 - P_{LZ})$ (constructive interference). Suppression of the transitions at the destructive interference is called the *coherent destruction of tunneling*. The constructive interference gives the probability twice as much as when the interference is not taken into account, as in Eq. (2.75). This situation is analogous to the processes in the known optical Mach–Zehnder interferometer.

2.4.4. *Multiple passage: interference*

For the case of the multiple passage of the quasi-crossing point, we have that after n half-periods, the evolution matrix will be defined by the n-th power of the following matrix

$$\Xi \equiv NU = \begin{pmatrix} -ir & -t \\ t & ir \end{pmatrix} \begin{pmatrix} e^{-i\varsigma} & 0 \\ 0 & e^{i\varsigma} \end{pmatrix} = \begin{pmatrix} \alpha & -\gamma^* \\ \gamma & \alpha^* \end{pmatrix},$$

$$\alpha = -ire^{-i\varsigma}, \quad \gamma = te^{-i\varsigma}. \tag{2.76}$$

In order to raise this matrix, it is convenient to make it diagonal first. For this, we have to find the unitary matrix

$$A = \begin{pmatrix} a & -b^* \\ b & a^* \end{pmatrix}, \quad AA^\dagger = 1, \quad |a|^2 + |b|^2 = 1, \tag{2.77}$$

such that $A\Xi A^\dagger = \begin{pmatrix} e^{i\phi} & 0 \\ 0 & e^{-i\phi} \end{pmatrix} \equiv B$, where ϕ is the desired value.

Consider the equation $\Xi = A^\dagger B A$ and obtain

$$\begin{cases} \alpha = |a|^2 e^{i\phi} + |b|^2 e^{-i\phi}, \\ \gamma = -2iab\sin\phi. \end{cases} \tag{2.78}$$

From the former equation we find ϕ: $\cos\phi = \operatorname{Re}\alpha$. So, we have

$$\Xi^n = A^\dagger B A \cdot A^\dagger B A \cdot \ldots = A^\dagger \begin{pmatrix} e^{in\phi} & 0 \\ 0 & e^{-in\phi} \end{pmatrix} A = \begin{pmatrix} u_{11} & -u_{21}^* \\ u_{21} & u_{11}^* \end{pmatrix}. \tag{2.79}$$

Here we simplify the obtained matrix elements, taking into account (2.78) (here we are interested only in u_{21}):

$$u_{11} = |a|^2 e^{in\phi} + |b|^2 e^{-in\phi} = \cos n\phi + i\sin n\phi\,(2|a|^2 - 1)$$

$$= \cos n\phi + i\sin n\phi\frac{\operatorname{Im}\alpha}{\sin\phi}, \tag{2.80}$$

$$u_{21} = -2iab\sin n\phi = \frac{\gamma}{\sin\phi}\sin n\phi.$$

The absolute value of the off-diagonal element gives, as above, the probability of finding the system on the upper level:

$$P_+(n) = |\gamma|^2\frac{\sin^2 n\phi}{\sin^2\phi}. \tag{2.81}$$

Note that $1 = |\alpha|^2 + |\gamma|^2 = (\operatorname{Re}\alpha)^2 + (\operatorname{Im}\alpha)^2 + |\gamma|^2$ and $\operatorname{Re}\alpha = \cos\phi$, from where it follows

$$\sin^2\phi = (\operatorname{Im}\alpha)^2 + |\gamma|^2 = t^2 + r^2\cos^2\zeta, \tag{2.82}$$

and we can rewrite Eq. (2.81):

$$P_+(n) = \frac{t^2}{t^2 + r^2\cos^2\zeta}\sin^2 n\phi = \frac{P_{\text{LZ}}}{\cos^2\zeta + P_{\text{LZ}}\sin^2\zeta}\sin^2 n\phi. \tag{2.83}$$

Here, the second multiplier describes the time dependence, where $n = [\omega t/\pi]$, and the first multiplier describes the oscillations amplitude.

(When we defined n, the brackets stand for the integer part, so that $[\omega t/\pi]$ equals to the number of half-periods.) The amplitude has the maximum at $\zeta = \frac{\pi}{2} + k\pi$ and minimum at $\zeta = k\pi$. This corresponds to the cases of constructive and destructive interference, respectively, for which we have

$$
\begin{aligned}
\zeta = \frac{\pi}{2} + k\pi : \quad & \sin^2 \phi = t^2 \Rightarrow \phi \approx \sqrt{P_{\mathrm{LZ}}}, \\
& P_+(t) = \sin^2\left(\sqrt{P_{\mathrm{LZ}}}\, n\right) ; \\
\zeta = k\pi : \quad & \sin^2 \phi = t^2 + r^2 = 1 \Rightarrow \phi = \pi/2, \\
& P_+(t) = P_{\mathrm{LZ}} \sin^2\left(\frac{\pi}{2}n\right) .
\end{aligned}
\tag{2.84}
$$

These formulas are illustrated in Fig. 2.8. We note the resemblance of the constructive-interference case with the Rabi oscillations. Indeed, if the driving frequency ω is large, then the Landau–Zener probabilities are small and the steps are little visible on the background of the large-scale oscillations.

* Remarkably, if we would not assume $\varepsilon_0 = 0$ and consider the case of the large-amplitude oscillations (of which the details can be

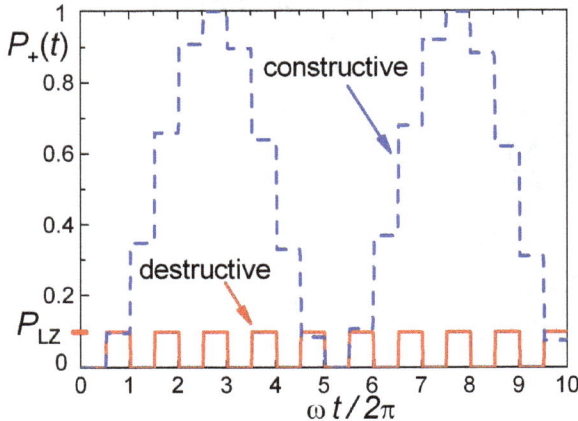

Fig. 2.8. Constructive (dashed line) and destructive (solid line) interference. The time dependence of the upper-level occupation P_+ is plotted for constructive ($\zeta_1 = \pi/2 + k\pi$) and destructive ($\zeta_1 = k\pi$) interference at $\varepsilon_0 = 0$ and $P_{\mathrm{LZ}} = 0.1$.

found in Footnote 17 on page 54), then the result of this approach, within the framework of the adiabatic-impulse model, would coincide with the formula (2.39), where $\Delta_k = \Delta\sqrt{\frac{2\hbar\omega}{\pi A}}\cos\left(\frac{A}{\hbar\omega} - \frac{\pi}{4}(2k+1)\right)$ and which correctly describes the Bessel function asymptote.

2.5. * Quantum relaxation theory

Above, in Section 2.3, we mentioned that the interaction with the environment leads to the relaxation, and we considered this for a two-level system, introducing the phenomenological relaxation times. Here we will pay more attention to the description of the interaction of a quantum system with the environment and we will consistently present the quantum relaxation theory. It is worth emphasizing that accounting for the unavoidable dissipative environment is particularly important in mesoscopic systems, which interact with the environment much more strongly than their microscopic counterparts. We will obtain the so-called Master kinetic equation, and we will show how in a particular case of a two-level system the Bloch equation can be obtained from this. And we strongly believe that it is very instructive to compare the two approaches which we use to obtain this equation, cf. the phenomenological approach above and the microscopic detailed derivation below. In this Section we will follow, to a large degree, the approach from [Blum 1981].

2.5.1. *Interaction of a quantum system with a reservoir*

Well, consider an open system S, interacting with a large reservoir (thermostat) R. The Hamiltonian for the system is

$$H = H_{\mathrm{S}} + H_{\mathrm{R}} + V \equiv H_0 + V, \qquad (2.85)$$

where we separated the part $H_0 = H_{\mathrm{S}} + H_{\mathrm{R}}$ responsible for the non-interacting system and the reservoir.

We change from the Schrödinger representation to the interaction representation. The density operator is transformed as follows

$$\rho_{\mathrm{I}} = e^{\frac{i}{\hbar}H_0 t}\rho\, e^{-\frac{i}{\hbar}H_0 t}. \qquad (2.86)$$

Analogously we obtain the interaction operator in the new representation V_I. The Liouville–von Neumann equation, $\dot{\rho} = -\frac{i}{\hbar}[H_0 + V, \rho]$, becomes

$$\dot{\rho}_I = -\frac{i}{\hbar}[V_I, \rho_I]. \qquad (2.87)$$

Its formal solution is

$$\rho(t)_I = \rho(0)_I - \frac{i}{\hbar}\int_0^t dt' [V(t')_I, \rho(t')_I]. \qquad (2.88)$$

Let us put this solution back into the Liouville–von Neumann equation (2.87) and take the trace of this equation over the unobservable reservoir degrees of freedom. Then, for the desired reduced density matrix of our system, $\rho_{S,I} = Tr_R\rho_I$, we obtain the equation

$$\dot{\rho}_{S,I} = -\frac{i}{\hbar}Tr_R[V(t)_I, \rho(0)_I] - \frac{1}{\hbar^2}\int_0^t dt' Tr_R[V(t)_I, [V(t')_I, \rho(t')_I]].$$
$$(2.89)$$

In order to simplify this integro-differential equation, usually one can make the following two key assumptions.

First, the dynamics of the system S does not influence the state of the "large" reservoir R. This allows the replacement below to be made:

$$\rho(t)_I \rightarrow \tilde{\rho}(t)_I = \rho(t)_{S,I}\rho(0)_R. \qquad (2.90)$$

This replacement is known as the *irreversibility condition*, since after this, the Liouville–von Neumann equation describes the irreversible relaxation of the system S. It is assumed that the reservoir is in thermal equilibrium and then its statistical operator has the known form:

$$\rho(0)_R = \frac{\exp(-H_R/k_B T)}{\Sigma}, \quad \Sigma = Tr_R \exp(-H_R/k_B T). \qquad (2.91)$$

The second assumption is the so-called *Markov approximation*: it is assumed that the interaction with the reservoir destroys the information about the system state in the past. This means that $\dot{\rho}_{S,I}$

depends only on the instantaneous value $\rho(t)_{\text{S,I}}$. Therefore, in the r.h.s. of Eq. (2.89) for $\dot{\rho}_{\text{S,I}}$ we make the following replacement

$$\rho(t')_{\text{S,I}} \to \rho(t)_{\text{S,I}}. \tag{2.92}$$

The Markov approximation will allow us to reduce the integro-differential equation to the system of linear differential equations with time-independent coefficients.

Thus, from Eq. (2.89) we obtain

$$\dot{\rho}_{\text{S,I}} = -\frac{i}{\hbar} Tr_{\text{R}}[V(t)_{\text{I}}, \rho(0)_{\text{S,I}}\, \rho(0)_{\text{R}}]$$

$$-\frac{1}{\hbar^2} \int_0^t dt'\, Tr_{\text{R}}[V(t)_{\text{I}}, [V(t')_{\text{I}}, \rho(t)_{\text{S,I}}\, \rho(0)_{\text{R}}]]. \tag{2.93}$$

2.5.2. *Temporal correlation functions*

For further transformations of Eq. (2.93), we assume that in the interaction operator, we can separate the operators corresponding to our quantum system (Q_i) and the operators corresponding to the reservoir (F_i), which means that we can write $V = \sum_i Q_i F_i$. Then in the interaction representation we have

$$V(t)_{\text{I}} = \sum_i Q(t)_i F(t)_i, \quad Q(t)_i = e^{\frac{i}{\hbar}H_{\text{S}}t} Q_i e^{-\frac{i}{\hbar}H_{\text{S}}t},$$

$$F(t)_i = e^{\frac{i}{\hbar}H_{\text{R}}t} F_i e^{-\frac{i}{\hbar}H_{\text{R}}t}. \tag{2.94}$$

We put this in Eq. (2.93), and then we have

$$\dot{\rho}(t)_{\text{S,I}} = -\frac{i}{\hbar} \sum_i Tr_{\text{R}}[Q(t)_i F(t)_i, \rho(0)_{\text{S,I}}\rho(0)_{\text{R}}]$$

$$-\frac{1}{\hbar^2} \sum_{i,j} \int_0^t dt'\, Tr_{\text{R}}[Q(t)_i F(t)_i, [Q(t')_j F(t')_j, \rho(t)_{\text{S,I}}\, \rho(0)_{\text{R}}]].$$

$$\tag{2.95}$$

We rewrite the commutators:

$$\dot{\rho}(t)_{S,I} =$$

$$-\frac{i}{\hbar}\sum_i \{Q(t)_i\rho(0)_{S,I}Tr_R(F(t)_i\rho(0)_R) - \rho(0)_{S,I}Q(t)_iTr_R(\rho(0)_RF(t)_i)\}$$

$$-\frac{1}{\hbar^2}\sum_{i,j}\int_0^t dt'\{(Q(t)_iQ(t')_j\rho(t)_{S,I} - Q(t')_j\rho(t)_{S,I}Q(t)_i)Tr_R(F(t)_iF(t')_j\rho(0)_R)$$

$$-(Q(t)_i\rho(t)_{S,I}Q(t')_j - \rho(t)_{S,I}Q(t')_jQ(t)_i)Tr_R(F(t')_jF(t)_i\rho(0)_R)\}. \qquad (2.96)$$

We can see that in this equation there are terms of two kinds: the mean values of operators F_i and the two-time mean values. Consider them by turns. Let us take the trace in the basis $\{|N\rangle\}$ of the eigenstates of the reservoir Hamiltonian H_R:

$$Tr_R(F(t)_i\rho(0)_R) \equiv \langle F(t)_i\rangle = \sum_N \langle N|F(t)_i\rho(0)_R|N\rangle$$

$$= \sum_N \langle N|F(t)_i|N\rangle\langle N|\rho(0)_R|N\rangle. \qquad (2.97)$$

Here, it was taken into account that $1 = \sum_{N'}|N'\rangle\langle N'|$ and that the density matrix $\rho_{N'N}$ in equilibrium is diagonal. We assume that in this representation, the operators F_i do not have diagonal elements, $\langle N|F(t)_i|N\rangle = 0$. (Otherwise, the Hamiltonian H_R could be redefined so as to satisfy this.) We obtain $\langle F(t)_i\rangle = 0$.

Consider next the two-time correlation functions

$$Tr_R(F(t)_iF(t')_j\rho(0)_R) \equiv \langle F(t)_iF(t')_j\rangle. \qquad (2.98)$$

It is assumed that the reservoir is large and that the effects of interaction are quickly damped, during the short time, known as the *reservoir correlation time* τ. To be more precise, this time is much shorter than the relaxation of our quantum system, which justifies the Markov approximation. This means that for a large time difference, $t - t' \gg \tau$, the correlator is equal to

$$\langle F(t)_iF(t')_j\rangle = \langle F(t)_i\rangle\langle F(t')_j\rangle = 0. \qquad (2.99)$$

One can see from the definition that the correlation functions are stationary, which means that they depend only on the difference $t-t'$

[see Eqs. (2.99) and (2.94)]:

$$\langle F(t)_i F(t')_j \rangle = \langle F(t-t')_i F_j \rangle. \tag{2.100}$$

Then, in Eq. (2.96) we can change the integration variable, $t'' = t - t'$, and, since for $t'' \gg \tau$ we have Eq. (2.99), we can change the upper limit of integration for ∞. We obtain

$$\dot{\rho}(t)_{\text{S,I}} = -\frac{1}{\hbar^2} \sum_{i,j} \int_0^\infty dt'' \left\{ [Q(t)_i, Q(t-t'')_j \rho(t)_{\text{S,I}}] \langle F(t'')_i F_j \rangle \right.$$

$$\left. - [Q(t)_i, \rho(t)_{\text{S,I}} Q(t-t'')_j] \langle F_j F(t'')_i \rangle \right\}. \tag{2.101}$$

We can see that all the information about the reservoir is now contained in the correlation functions.

2.5.3. *Master equation*

Let us change from operators to matrix elements — in the eigenbasis $\{|m\rangle\}$ of the Hamiltonian H_{S}. From the definition (2.94) we have

$$\langle m|Q(t)_i|n \rangle = e^{i\omega_{mn}t} \langle m|Q_i|n \rangle,$$

$$\hbar\omega_{mn} = E_m - E_n. \tag{2.102}$$

We next define the tensors

$$\Gamma^+_{mkln} \triangleq \frac{1}{\hbar^2} \sum_{ij} \langle m|Q_i|k \rangle \langle l|Q_j|n \rangle \int_0^\infty dt'' e^{-i\omega_{ln}t''} \langle F(t'')_i F_j \rangle,$$

$$\tag{2.103}$$

$$\Gamma^-_{mkln} \triangleq \frac{1}{\hbar^2} \sum_{ij} \langle m|Q_j|k \rangle \langle l|Q_i|n \rangle \int_0^\infty dt'' e^{-i\omega_{mk}t''} \langle F_j F(t'')_i \rangle.$$

$$\tag{2.104}$$

From this definition, and taking into account the Hermiticity of the operators, it follows that

$$\Gamma^-_{nlkm} = (\Gamma^+_{mkln})^*. \tag{2.105}$$

Then we obtain the following (and it is recommended to the reader to make sure of this, accounting for the condition that $1 = \sum_n |n\rangle\langle n|$):

$$\langle m'|\dot{\rho}(t)_{\text{S,I}}|m\rangle = \sum_{n',n} \langle n'|\rho(t)_{\text{S,I}}|n\rangle R_{m'mn'n} e^{i(\omega_{m'n'}-\omega_{mn})t}. \quad (2.106)$$

Here, we defined the tensor

$$R_{m'mn'n} = -\sum_k \delta_{mn}\Gamma^+_{m'kkn'} + \Gamma^+_{nmm'n'} + \Gamma^-_{nmm'n'}$$

$$-\sum_k \delta_{n'm'}\Gamma^-_{nkkm}, \quad (2.107)$$

known as the *Redfield* (or Bloch–Redfield) *tensor*. This can be conveniently split into two parts:

$$R_{m'mn'n} =: W_{mn}\delta_{n'n} - \gamma_{m'm}\delta_{m'n'}\delta_{mn},$$

$$W_{mn} \triangleq \Gamma^+_{nmmn} + \Gamma^-_{nmmn} = 2\text{Re}\Gamma^+_{nmmn}, \quad (2.108)$$

$$\gamma_{m'm} \triangleq \sum_k (\Gamma^+_{m'kkm'} + \Gamma^-_{mkkm}) - \Gamma^+_{mmm'm'} - \Gamma^-_{mmm'm'}.$$

We now make the next, so-called, *secular approximation*, where it is assumed that in the sum in Eq. (2.106) the main contribution comes from the time-independent terms, for which $\omega_{m'n'} - \omega_{mn} = 0$. This approximation assumes that we are interested in the "coarse-grained" derivative. In other words, the procedure involving averaging of the equation over small time intervals is assumed, during which the derivative changes little and the temporal terms of the form $e^{i\omega t}$ change fast. This is analogous to the rotating-wave approximation, considered above. The equation $\omega_{m'n'} - \omega_{mn} = 0$ is fulfilled in one of the three cases:

$$\begin{cases} (1) \ m' = n', \ m = n, \ m' \neq m; \\ (2) \ m' = m, \ n' = n, \ m' \neq n'; \\ (3) \ m' = m = n' = n. \end{cases} \quad (2.109)$$

Writing down the corresponding terms, we obtain (and it is recommended that the reader checks this)

$$\langle m'|\dot{\rho}(t)_{\text{S,I}}|m\rangle = \delta_{m'm} \sum_{n\neq m} W_{mn}\langle n|\rho(t)_{\text{S,I}}|n\rangle - \gamma_{m'm}\langle m'|\rho(t)_{\text{S,I}}|m\rangle.$$

(2.110)

Let us write down this equation in the Schrödinger representation, by using $\rho_{\text{S,I}} = e^{\frac{i}{\hbar}H_{\text{S}}t}\rho_{\text{S}}e^{-\frac{i}{\hbar}H_{\text{S}}t}$,

$$\langle m'|\dot{\rho}(t)_{\text{S}}|m\rangle = -\frac{i}{\hbar}\langle m'|[H_{\text{S}},\rho(t)_{\text{S}}]|m\rangle + \delta_{m'm}\sum_{n\neq m} W_{mn}\langle n|\rho(t)_{\text{S}}|n\rangle$$

$$- \gamma_{m'm}\langle m'|\rho(t)_{\text{S}}|m\rangle.$$

(2.111)

We rewrite this more compactly, also omitting the subscript S,

$$\dot{\rho}_{m'm} = -\frac{i}{\hbar}[H,\rho]_{m'm} + \delta_{m'm}\sum_{n\neq m} W_{mn}\rho_{nn} - \gamma_{m'm}\rho_{m'm}.$$ (2.112)

This is the so-called *generalized Master equation*. This equation is also known as the *Bloch–Redfield equation*. In practice, as the next step, we should calculate the tensors Γ_{lmnk} for a specific case, setting a physically grounded model for the reservoir and its interaction with the system, so as to obtain W_{mn} and $\gamma_{m'm}$. Let us make several general remarks concerning the Master equation and then consider in detail the important and illustrative case of a two-level system.

In particular, for the diagonal elements (see definition of the tensor $\gamma_{m'm}$ for $m' = m$):

$$\dot{\rho}_{mm} = \sum_{n\neq m} W_{mn}\rho_{nn} - \rho_{mm}\sum_{n\neq m} W_{nm}.$$ (2.113)

Here the first term defines the gain and the second term defines the diminution of the population of the level m. The value W_{mn} holds the meaning of the transition rate, which means that this is the *transition probability* in unit time from the level n to the level m.

These transitions are induced by the interaction of our system with the reservoir. Equation (2.113) is called the *Pauli principal kinetic equation.*

Consider the transition rates $W_{mn} = 2\mathrm{Re}\Gamma^+_{nmmn}$ in detail. For calculations we use the eigen-basis of the reservoir $\{|N\rangle\}$:

$$W_{mn} = \frac{2}{\hbar^2}\mathrm{Re}\langle n|Q_i|m\rangle\langle m|Q_j|n\rangle \int_0^\infty dt e^{-i\omega_{mn}t}$$
$$\times \sum_{N,N',N''} \langle N|F(t)_i|N'\rangle\langle N'|F_j|N''\rangle\langle N''|\rho(0)_\mathrm{R}|N\rangle.$$

$$(2.114)$$

Here the summation over i, j is implied. Now, remember the definition $F(t)_i = e^{\frac{i}{\hbar}H_\mathrm{R}t}F_i e^{-\frac{i}{\hbar}H_\mathrm{R}t}$ and take into account that $\rho(0)_\mathrm{R}$ is the diagonal matrix, $\langle N|\rho(0)_\mathrm{R}|N\rangle = \exp(-E_N/k_\mathrm{B}T)/\Sigma$, then we obtain

$$W_{mn} = \frac{2}{\hbar^2\Sigma}\sum_{N,N'}\mathrm{Re}\underbrace{\langle nN|Q_iF_i|mN'\rangle\langle mN'|Q_jF_j|nN\rangle}_{|\langle nN|V|mN'\rangle|^2}e^{-E_N/k_\mathrm{B}T}$$
$$\times \int_0^\infty dt e^{\frac{i}{\hbar}(E_N-E_{N'}-E_m+E_n)t}.$$

$$(2.115)$$

Here we took into consideration the Hermiticity of the interaction operator. From this operator, apropos, it follows the equality of the probabilities for the direct $(mN \to nN')$ and inverse transitions: $|\langle mN|V|nN'\rangle|^2 = |\langle nN'|V|mN\rangle|^2$. We further note that

$$2\mathrm{Re}\int_0^\infty dt e^{i\omega t} = 2\int_0^\infty dt\cos\omega t \overset{\triangle}{=} \int_{-\infty}^\infty dt e^{i\omega t} = 2\pi\delta(\omega) \quad (2.116)$$

and obtain

$$W_{mn} = \frac{2\pi}{\hbar\Sigma}\sum_{N,N'}|\langle nN|V|mN'\rangle|^2 e^{-E_N/k_\mathrm{B}T}\delta(E_N - E_{N'} - E_m + E_n).$$

$$(2.117)$$

Here the first multiplier describes the transition probability for the system from the state $|m\rangle$ into the state $|n\rangle$ and for the reservoir from the state $|N'\rangle$ into $|N\rangle$. This formula is called the *Fermi's golden rule* for the transition rates.

For the opposite transition we have

$$W_{nm} = \frac{2\pi}{\hbar\Sigma} \sum_{N,N'} |\langle mN|V|nN'\rangle|^2 e^{-E_N/k_\mathrm{B}T} \delta(E_N - E_{N'} - E_n + E_m)$$

$$= \|N' \leftrightarrow N\| = \frac{2\pi}{\hbar\Sigma} \sum_{N,N'} |\langle mN'|V|nN\rangle|^2 \underbrace{e^{-E_{N'}/k_\mathrm{B}T}}_{E_{N'}=E_N+E_n-E_m}$$

$$\times \delta(E_{N'} - E_N - E_n + E_m) = e^{(E_m - E_n)/k_\mathrm{B}T} W_{mn}. \qquad (2.118)$$

Here we used the hermiticity of the operator V and the evenness of the δ-function. We obtain the relation between the transition rates upwards and downwards:

$$\frac{W_{mn}}{W_{nm}} = \exp\left(-\frac{E_m - E_n}{k_\mathrm{B}T}\right). \qquad (2.119)$$

2.5.4. *The case of a two-level system*

Consider further the Master equation (2.112) for a two-level system. For the diagonal elements we have

$$\dot{\rho}_{mm} = -\frac{i}{\hbar}[H,\rho]_{mm} + W_{mn}\rho_{nn} - W_{nm}\rho_{mm}, \quad n \neq m. \qquad (2.120)$$

From the normalization condition $\rho_{00} + \rho_{11} = 1$ it follows that $\dot{\rho}_{11} = -\dot{\rho}_{00}$. Then we rewrite Eqs. (2.120) and (2.119):

$$\dot{\rho}_{00} = -\frac{i}{\hbar}[H,\rho]_{00} + W_{01}\rho_{11} - W_{10}\rho_{00}, \qquad (2.121)$$

$$\frac{W_{10}}{W_{01}} = \exp\left(-\frac{E_1 - E_0}{k_\mathrm{B}T}\right) = \exp\left(-\frac{\Delta E}{k_\mathrm{B}T}\right). \qquad (2.122)$$

From these two equations, in thermal equilibrium, it follows that

$$\frac{\rho_{11}^{(0)}}{\rho_{00}^{(0)}} = \frac{W_{10}}{W_{01}} = \exp\left(-\frac{\Delta E}{k_\mathrm{B}T}\right). \qquad (2.123)$$

This means that in thermal equilibrium we have the Boltzmann distribution.

We now exclude ρ_{11} and W_{mn} from Eq. (2.121). First, consider

$$W_{01}\rho_{11} - W_{10}\rho_{00} = W_{01} - \rho_{00}(W_{01} + W_{10}). \tag{2.124}$$

Introduce the definition

$$\frac{1}{T_1} \stackrel{\triangle}{=} W_{01} + W_{10}. \tag{2.125}$$

We obtain $\rho_{00}^{(0)}$ from Eq. (2.123):

$$\frac{1 - \rho_{00}^{(0)}}{\rho_{00}^{(0)}} = \frac{1}{\rho_{00}^{(0)}} - 1 = \frac{W_{10}}{W_{01}} \quad \Rightarrow \quad \rho_{00}^{(0)} = \frac{W_{01}}{W_{01} + W_{10}} = T_1 W_{01}. \tag{2.126}$$

Let us express from here $W_{01} = \rho_{00}^{(0)}/T_1$, then from Eqs. (2.124) and (2.121) we have the equation for ρ_{00}

$$\dot{\rho}_{00} = -\frac{i}{\hbar}[H, \rho]_{00} - \frac{\rho_{00} - \rho_{00}^{(0)}}{T_1}. \tag{2.127}$$

For the off-diagonal matrix elements, after introducing the notation

$$\frac{1}{T_2} \stackrel{\triangle}{=} \gamma_{01}, \tag{2.128}$$

from Eq. (2.112) we obtain the second of the Bloch equations:

$$\dot{\rho}_{01} = -\frac{i}{\hbar}[H, \rho]_{01} - \frac{\rho_{01}}{T_2}. \tag{2.129}$$

Thus, in the case of a two-level system, we have derived the Bloch equations from the generalized Master equation. These are the very same equations which we have obtained in §2.3 phenomenologically. So, the reader can compare these two very different theoretical approaches to the description of an open quantum system.

2.5.5. *Lindblad equation*

An alternative to the approach described above is the Master equation in the Lindblad form, or simply, the Lindblad equation. We will write it down, without derivation, from the textbook.[20]

---

[20]F. Laloë, Do we really understand Quantum Mechanics? Cambridge, UK: Cambridge University Press (2012), §6.4.

We specify that the results, obtained from the solution of either the Lindblad equation or the Bloch–Redfield equations can differ in the general case; refer to this in Footnote 21. However, for the simplest system, a qubit, they give an identical result. So, here we will write down the Lindblad equation for the general case and we will consider the particular case of a single qubit.

Quantum dynamics of an open system with the Hamiltonian H is described by the Lindblad equation:

$$\dot{\rho} = -\frac{i}{\hbar}[H, \rho] + \sum_\alpha \check{L}_\alpha[\rho], \qquad (2.130)$$

where the interaction with the environment is described by the *Lindblad superoperators*

$$\check{L}_\alpha[\rho] = L_\alpha \rho L_\alpha^\dagger - \frac{1}{2}\{L_\alpha^\dagger L_\alpha, \rho\} = \frac{1}{2}[L_\alpha \rho, L_\alpha^\dagger] + \frac{1}{2}[L_\alpha, \rho L_\alpha^\dagger]. \qquad (2.131)$$

Here we have given two equivalent definitions. The interaction with the environment results in relaxation; different channels of relaxation are enumerated by the subscripts α and are described by the *Lindblad operators* L_α, which have the general form as the following, $L_\alpha = \sqrt{\Gamma_\alpha} A_\alpha$.

In particular, for a qubit $L_{\text{relax}} = \sqrt{\Gamma_1}\sigma$ and $L_\phi = \sqrt{\Gamma_\phi/2}\sigma_z$. Instead of the relaxation rates, it is often more convenient to use respective times: $T_1 = \Gamma_1^{-1}$ and $T_\phi = \Gamma_\phi^{-1}$. Then for the two relaxation channels, we have:

$$\check{L}_{\text{relax}}[\rho] = \frac{1}{2T_1}(2\sigma\rho\sigma^\dagger - \{\sigma^\dagger\sigma, \rho\}), \quad \check{L}_\phi[\rho] = \frac{1}{2T_\phi}(\sigma_z\rho\sigma_z - \rho). \qquad (2.132)$$

Now, it is convenient to take the density matrix in the form $\rho = \frac{1}{2}\begin{pmatrix} 1+Z & X-iY \\ X+iY & 1-Z \end{pmatrix}$. And then the reader only needs to multiply the respective matrices in Eq. (2.132), group the obtained result with

---

[21] J.R. Johansson, P.D. Nation, and F. Nori, QuTiP 2: A Python framework for the dynamics of open quantum systems, Comp. Phys. Comm. **184**, 1234 (2013).

the Pauli matrices, and obtain, instead of Eq. (2.130), the following

$$\frac{1}{2}(\dot{X}\sigma_x + \dot{Y}\sigma_y + \dot{Z}\sigma_z)$$

$$= -\frac{i}{\hbar}[H, \rho] - \frac{1}{2T_1}(1 + Z)\sigma_z - \frac{1}{2T_2}X\sigma_x - \frac{1}{2T_2}Y\sigma_y,$$

$$(2.133)$$

$$\frac{1}{T_2} \triangleq \frac{1}{T_\phi} + \frac{1}{2T_1}. \qquad (2.134)$$

The obtained equation coincides with the Bloch equations considered above. Here the relaxation of the diagonal components of the density matrices (the populations) appears during the time of energy relaxation T_1, and the relaxation of the off-diagonal elements (coherences) appears during the decoherence time T_2.

Conclusion to Chapter 2

A two-level system is, probably, the most fundamental model in non-relativistic quantum mechanics. (Another fundamental system is the harmonic oscillator, and to this we will devote a significant part of Chapter 5.) On the other hand, we have seen in the previous Chapter that the modern quantum technologies are based on this two-level system, a qubit. That is why we have paid so much attention to the dynamics of a qubit.

The detailed approach allowed us to consider the number of notions in this Chapter, to which little attention is paid in traditional textbooks of quantum mechanics, but which are frequently met in different areas of modern physics. We here have learned such notions as the spectrum of a two-level system, Rabi oscillations, pure and mixed states, Landau–Zener transitions and Stückelberg oscillations, equations for a density matrix — the Liouville–von Neumann equation, the Bloch–Redfield equation and the Lindblad equation.

For the description of the dynamics of a qubit driven by a periodic signal, we have used three different approaches. The first regime was the theory of periodic perturbation for weak amplitudes $A \ll \Delta$

(which resulted in the Rabi oscillations), the second regime was for the strong and fast excitation, $A \cdot \hbar\omega/\Delta^2 \gg 1$, for which we have used the rotating wave approximation, and the third regime was related to the adiabatically slow changes, when the adiabaticity parameter was large, $\frac{\Delta^2}{A \cdot \hbar\omega} \geq 1$, that is $A \cdot \hbar\omega/\Delta^2 \leq 1$, for which we have used the so-called adiabatically-impulse method, also known as the transfer-matrix method. At that, each of these regimes, strictly speaking, was correct for the mentioned inequalities. It is remarkable, however, that the regions of their applicability are much wider and are essentially overlapping. This can be checked by comparing the analytic solutions with the numerical ones, as e.g. in Footnote 22. And in this Chapter here, we have demonstrated the mutual agreement of all these methods by considering respective limiting cases.

Problems for independent work and for self-assessment

2.1. (**) Obtain the qubit energies and eigenstates from its Schrödinger equation.

2.2. (**) Do the same by means of the transformation $S = \exp(i\zeta\sigma_y/2)$.

2.3. (***) Describe the Rabi oscillations for resonant driving, with $\omega = \omega_q$.

2.4. (****) Describe the Rabi oscillations for nonzero frequency detuning, $\delta\omega = \omega - \omega_q \neq 0$.

2.5. (*) Based on Eqs. (2.29)–(2.31), describe the spectroscopy of a driven two-level system.

2.6. (****) Making use of the respective rotating-wave approximation, describe the multi-photon excitation of a strongly driven two-level system, i.e. obtain Eq. (2.39).

2.7. (**) Rewrite the Liouville–von Neumann equation, with the pseudospin Hamiltonian, in the form of the Bloch equation, $\dot{\vec{R}} = \vec{R} \times \vec{H}$.

---

[22]S. Ashhab, J.R. Johansson, A.M. Zagoskin, and F. Nori, Two-level systems driven by large-amplitude fields, Phys. Rev. A **75**, 063414 (2007).

2.8. (*) Show that the free evolution of a spin is described by the quantum beats.

2.9. (****) Solve the Bloch equations with strong driving and with dissipation and obtain the formula, describing the multi-photon resonances, Eq. (2.57).

2.10. (***) Given the formula (2.57), plot the upper-level occupation probability as a function of the bias and driving amplitude, and analyze its usefulness for the interferometry.

2.11. (**) In analogy with the tasks 2.1–2.2, obtain the adiabatic energy levels and draw them.

2.12. (**) Describe the adiabatic evolution, by introducing the transfer matrix.

2.13. (**) Given the quasi-classical formula for the transition (2.64), obtain the Landau–Zener formula.

2.14. (***) Solve the respective Schrödinger equation and be convinced of the accuracy of the Landau–Zener formula.

2.15. (*****) Matching the solutions of the Schrödinger equation close to the avoided-level crossing and far from it, obtain the transfer matrix N for the Landau–Zener transition (see Appendix in Footnote 17 on page 54). Note that we need this result, Eq. (2.68), to describe the interference.

2.16. (**) Multiply the transfer matrices for a one-period evolution and obtain the Stückelberg oscillations, Eq. (2.74).

2.17. (***) Obtain the upper-level occupation after n half-periods, Eq. (2.83).

2.18. (**) Having obtained Eq. (2.83), demonstrate the cases of constructive and destructive interference.

2.19. (*****) Following the lines of Sec. 2.5, describe the relaxation theory and obtain the Master equation (2.112). This is admittedly a demanding task, but would be useful especially for students specializing in theoretical physics to attempt. Note that this is not necessary for understanding the main ideas of this course, but students studying quantum technologies are also welcome to try this.

2.20. (**) Verify Eq. (2.105).

2.21. (***) Demonstrate that Eq. (2.110) follows from Eq. (2.106) in the secular approximation.

2.22. (**) Having the Master equation (2.112), analyze it for the diagonal elements (occupations).

2.23. (***) Obtain the Bloch equations from the Master equation (2.112) for a qubit.

2.24. (***) Obtain the Bloch equations from the Lindblad equation (2.130) for a qubit.

Chapter 3

SUPERCONDUCTING QUANTUM CIRCUITS

> "... a river is simultaneously in different places
> ... everywhere at the same time ... for it, only
> the present exists ..."
>
> H. Hesse[23]

In accordance with the aims of our course, it is convenient to speak about a specific system. And here, appropriate examples are the qubits. Indeed, as discussed above, we note that (a) the qubits present the basic system of quantum mechanics, considering which, we can expound many basic problems, (b) they are interesting for possible applications in the field of quantum engineering, and (c) the possibility of these applications, in turn, is the locomotive of fundamental research in mesoscopic, superconducting and semiconducting systems.

One of the requirements for a physical realization of qubits is the *integrability*, which is the possibility of wiring up qubits with each other and with the controlling electronics. Such a possibility can be represented by solid-state realizations. For the least level of dissipation, superconductors are used.

The simplest possibility of the integrable electric circuit is represented by an electric resonant LC circuit. Consider the example: a circuit of two elements connected in parallel, an inductance L and a

---

[23]H. Hesse, Siddhartha, SPb: Azbuka (2007) {translated from Russian by the author}.

capacitance C. The voltage drop on them is equal to $q/C = -L\dot{I}$, where q is the charge on the capacitor plate. The derivative from this equation gives $L\ddot{I} + I/C = 0$. And the structure of this equation, apart from the notations, coincides with the equation of a harmonic oscillator, $m\ddot{x} + kx = 0$. As it is known, after quantization with the harmonic potential we obtain equidistant energy levels. This does not allow the lowest two levels to be isolated. One then needs the nonlinearity in the system. This means that we need a nonlinear inductance, defining the link between the current derivative and the voltage, $V = -L\dot{I}$ with $L = L(I, V)$.

The only non-dissipative nonlinear element is a Josephson junction. In what follows, we will consider how such a junction can be described as a nonlinear inductance. After that, the principal types of superconducting qubits will be presented in detail. We will consider classical circuits and the procedure of their quantization.

Superconducting qubits can be realized on the basis of Josephson junctions.[24,25] Significant progress was reached in the study of such qubits: both single and coupled qubits were realized, superposition and entangled states were observed, the series of effects was studied in relation to transitions in such multi-level qubit devices, etc.[26] The first public and commercial quantum computers are built on superconducting qubits.[27]

3.1. Some information on superconductivity

Before describing the Josephson effect, consider briefly the basic properties of superconductors. For further acquaintance with such fascinating phenomena like superconductivity, we recommend the

[24]Yu. Makhlin, G. Schön, and A. Shnirman, Quantum-state engineering with Josephson-junction devices, Rev. Mod. Phys. **73**, 357 (2001).

[25]J. Clarke and F. K. Wilhelm, Superconducting quantum bits, Nature **453**, 1031 (2008).

[26]J. Q. You and F. Nori, Atomic physics and quantum optics using superconducting circuits, Nature **474**, 589 (2011); X. Gu, A.F. Kockum, A. Miranowicz, Y.X. Liu, and F. Nori, Microwave photonics with superconducting quantum circuits, Phys. Rep. **718–719**, pp. 1–102 (2017).

[27]See https://www.research.ibm.com/ibm-q and https://www.dwavesys.com.

textbook [Schmidt 1997]. For the first familiarization, one will find it useful to pick up some more information from other textbooks, such as Il'ichev and Greenberg.[28]

(1) *Infinite conductivity.* The major property of superconductors is the flow of dissipationless current. The current can be induced, for example, in a ring of the area S and inductance L, by putting it in the magnetic field B and decreasing the temperature. Then in the ring, the current appears, $I = \Phi/L = BS/L$. This assumes that its resistance equates to *zero* when the temperature is below the *critical temperature* T_c. The characteristic value for the maximal, or as they say, critical, current density is 10^6 A/cm^2. Correspondingly, the heat liberation is equal to zero, according to the Joule–Lenz law, $Q = I^2 R t$. For the technologically important aluminum and niobium the critical temperatures are $T_c = 1.2$ and 9.2 K, respectively. The boiling temperature of liquid helium is 4.2 K and that of liquid nitrogen is 77 K. In this respect, the discovery of the so-called high-temperature superconductors (HTSC) with $T_c > 77$ K such as yttrium ceramics ($T_c \sim 90$ K) was important. Interestingly, to date, there is no generally accepted theory explaining the superconductivity in HTSC.

(2) *Macroscopic wave function of the condensate.* The conventional superconductivity is explained within the Bardeen–Cooper–Schrieffer theory by coupling electrons, which form the so-called *Cooper pairs* with zero spin. The size of a Cooper pair is characterized by the *coherence length* ξ_0. The aggregate of such bosons forms the Bose–Einstein condensate, which means that the electron pairs are situated on the same quantum level and they can be described by a single wave function, $\Psi(\mathbf{r}) = |\Psi(\mathbf{r})|e^{i\varphi(\mathbf{r})}$. The wave function is normalized by the density of the Cooper pairs, $|\Psi(\mathbf{r})| = \sqrt{n_s}$. For the current density of the Cooper pairs with charge $2e$ and mass $2m$ (where we denoted, with m and

---

[28]E. V. Il'ichev and Ya. S. Greenberg, Quantum informatics and quantum bits on the base of superconducting and Josephson structures, Novosibirsk, NSTU (2013).

$e = -|e|$, the mass and the charge of an electron) we have

$$\mathbf{j}_s = (2e)\left(\Psi^* \frac{\mathbf{p} - 2e\mathbf{A}/c}{2(2m)}\Psi + c.c.\right)$$

$$= -i\frac{e\hbar}{2m}(\Psi^*\nabla\Psi - \Psi\nabla\Psi^*) - \frac{2e^2}{mc}|\Psi|^2\mathbf{A} \qquad (3.1)$$

$$= 2e|\Psi|^2\frac{\hbar\nabla\varphi - 2e\mathbf{A}/c}{2m} \equiv 2en_s\mathbf{v}_s.$$

This means that the phase gradient $\nabla\varphi$ defines the current in the system via the *superfluid velocity* $\mathbf{v}_s$. Here it was taken into account that the presence of a magnetic field corresponds to the replacement $\mathbf{p} \to \mathbf{p} - (2e)\mathbf{A}/c$ and $\mathbf{p} = -i\hbar\nabla$. So, the supercurrent is related to the phase gradient as follows:

$$\mathbf{j}_s = 2en_s\frac{\hbar\nabla\varphi - 2e\mathbf{A}/c}{2m}. \qquad (3.2)$$

(3) *Perfect diamagnetism.* A magnetic field does not penetrate into bulk superconductors, unlike normal metals, but rather is expelled from them. This is related to the appearance of the surface current, of which the field shields the external magnetic field, known as the *Meissner effect*. In order to be convinced in this, consider a homogenous superconductor in a weak magnetic field (such that we may assume $n_s = const$) and in the thermodynamic equilibrium (when there is no normal current, i.e. $\mathbf{j} = \mathbf{j}_s$). Then from Eq. (3.2) we obtain

$$\operatorname{rot}\mathbf{j} = -\frac{2e^2n_s}{mc}\mathbf{B}. \qquad (3.3)$$

This is the *London equation*. To this, we have to add the Maxwell equations, $\operatorname{rot}\mathbf{B} = \frac{4\pi}{c}\mathbf{j}$ and $\operatorname{div}\mathbf{B} = 0$. Then, making use of the known relation, $\operatorname{rot}\operatorname{rot}\mathbf{B} = \nabla(\operatorname{div}\mathbf{B}) - \Delta\mathbf{B}$, we get

$$\operatorname{rot}\mathbf{j} = \frac{c}{4\pi}\operatorname{rot}\operatorname{rot}\mathbf{B} = -\frac{c}{4\pi}\Delta\mathbf{B} \overset{\wedge}{=} -\frac{2e^2n_s}{mc}\mathbf{B}$$

$$\Rightarrow \Delta\mathbf{B} = \frac{1}{\lambda^2}\mathbf{B}, \qquad (3.4)$$

where $\lambda^2 = \frac{mc^2}{8\pi e^2 n_s}$ is the so-called *field-penetration depth*. Indeed, for the flat boundary, we obtain the solution $\mathbf{B} = \mathbf{B}_0 e^{-x/\lambda}$. This means that the magnetic field penetrates into a superconductor only to the depth of the order of λ. This value is small, usually of the order of $0.1~\mu m$.

(4) *Magnetic flux quantization.* Consider a doubly connected super-conductor; a ring, for example. Let us mentally picture the circuit C inside the ring and integrate along it the formula (3.2). Then the l.h.s. gives zero, since inside the superconductor there is no magnetic field, and we have $0 = \int_C d\mathbf{l}(\hbar\nabla\varphi - 2e\mathbf{A}/c)$. Let us take into account that according to the Stokes theorem, the vector potential circulation gives the magnetic flux inside the ring, $\int_C d\mathbf{l}\mathbf{A} = \int_{S_C} d\mathbf{S}\nabla\times\mathbf{A} = \mathbf{B}\int_S d\mathbf{S} = BS = \Phi$. Here we assume that the magnetic field is homogeneous inside the ring, over the area S. We obtain $\Phi = \frac{\hbar c}{2e}\int_C d\mathbf{l}\nabla\varphi$. On the other hand, the requirement of the wave function to be single-valued gives that over the full path-tracing $\int_C d\mathbf{l}\nabla\varphi = \delta\varphi = -2\pi n$, where n is an integer and the negative sign is introduced for convenience. We obtain

$$\Phi = n\Phi_0, \quad \Phi_0 = \frac{hc}{2\,|e|}. \tag{3.5}$$

This means that the magnetic flux trapped by the supercon-ducting ring, $\Phi = BS$, can take only values multiple of the *flux quantum* Φ_0 ($\Phi_0 \approx 2\cdot 10^{-15}$ Wb). Note also that in a general case, the change of a phase is defined by the magnetic flux: $\delta\varphi = -2\pi\Phi/\Phi_0$.

(5) *Quasiparticle excitations.* In the non-stationary regime, the response of a superconductor is defined both by the superconduc-tive condensate and the quasiparticle excitations over the ground state. The excitations are separated by the *gap* $2\Delta_S$ from the ground state, and therefore they can be neglected at sufficiently low temperature, $k_B T \ll \Delta_S$. The value $2\Delta_S$ corresponds to the energy, which is necessary for splitting one Cooper pair. Another condition on the absence of the quasiparticle excitations is the absence of high-frequency fields with $\hbar\omega \sim \Delta_S$, which means that it is necessary also to have $\omega \ll \Delta_S/\hbar$.

3.2. Josephson effect

Consider a weak link between two bulk pieces of a superconductor — the banks of the junction — so that this junction does not significantly change the state of the superconductors. The role of a weak link can be played by, for instance, the thin interlayer of an insulator, as shown in Fig. 3.1. The characteristic value of the critical current for such junctions is 10 A/cm^2, which is much smaller than the critical current of the bulk superconductors. The presence of the perturbation in the form of the weak link leads to the interference of the banks' wave functions; as a result, the common wave function is formed. As we will see, this state is characterized by flow of the dissipationless current through the interlayer insulator. This is the *stationary Josephson effect*. Moreover, if the dc voltage is applied to the junction, then the alternating current would flow through the junction — the phenomena known as the *non-stationary Josephson effect*. Such current-carrying states are characterized by the phases of the condensate wave functions $\varphi_{1,2}$. Importantly, the observation of the Josephson effects allowed, for the first time in the history of physics, the quantum-mechanical values, which are the phases $\varphi_{1,2}$, to be directly linked to the macroscopic values — the current and the voltage.

Let us derive the formulas for the stationary and non-stationary Josephson effects in the simplest approach, proposed by Feynman (see Footnote 10 on page 16 and [Schmidt 1997]). Consider a weak link of two identical superconducting condensates in the

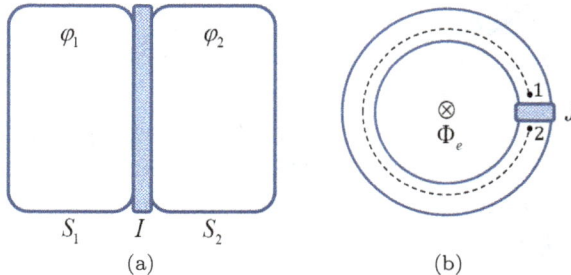

Fig. 3.1. (a) Josephson tunnel junction — two superconductors $S_{1,2}$, separated by a thin layer of an insulator I. (b) Josephson junction J, included in a ring, which is pierced by an external magnetic flux Φ_e.

junction banks, as in Fig. 3.1(a), which are described by the space-homogeneous condensate wave functions $\Psi_{1,2} = \sqrt{n_{s,1,2}}\, e^{i\varphi_{1,2}}$. It is assumed that the phase difference changes rapidly on the weak link and that this phase gradient — or the phase difference $\varphi = \varphi_1 - \varphi_2$ on the tunnel junction — parameterizes the current state. The current is assumed to be weak, that is why we can neglect its intrinsic magnetic field.

Consider a Josephson junction embedded in a ring as shown in Fig. 3.1(b). Let us integrate Eq. (3.2) along the circuit C, from point 1 to point 2. Since there is no supercurrent within a bulk superconductor, $\mathbf{j}_s = 0$, we have

$$\hbar \int_1^2 d\mathbf{l}\nabla\varphi = \hbar\left(\varphi_2 - \varphi_1\right) \equiv -\hbar\varphi \stackrel{\wedge}{=} \frac{2e}{c}\int_1^2 d\mathbf{l}\mathbf{A}$$

$$\approx \frac{2e}{c}\oint d\mathbf{l}\mathbf{A} = \frac{2e}{c}\Phi_e. \tag{3.6}$$

Here we made the approximation that the circuit was complemented up to the circular one, which is justified by the fact that the vector potential $\mathbf{A}$ does not have any singularity in the vicinity of the junction J. We have obtained the important relation that the phase difference on the junction is defined by the magnetic flux,

$$\varphi = 2\pi\frac{\Phi_e}{\Phi_0}. \tag{3.7}$$

So, we expect a Josephson structure to be highly sensitive to changes in a magnetic flux. We will discuss such magnetometers (so-called superconducting quantum interference devices, or SQUIDs) in more detail later.

Assume now that the electric potential difference (the voltage) V is applied to the junction (as in Fig. 3.1(a)) for convenience we assume $V < 0$, so that $eV > 0$. The two states of the junction banks, shown in Fig. 3.2, form a two-level system with the energy levels defined by the shift of the chemical potential and equal to $\pm eV$. Let us define the vector-states for the junction banks so that they form the basis of a two-level system

$$|\psi_1\rangle = \begin{pmatrix} 1 \\ 0 \end{pmatrix}, \quad |\psi_2\rangle = \begin{pmatrix} 0 \\ 1 \end{pmatrix}. \tag{3.8}$$

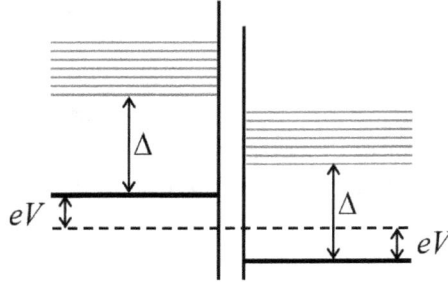

Fig. 3.2.   Energy diagram of a tunnel Josephson junction.

(Note that again we have a two-level system, though in a different context.) Let us expand the wave function in this basis:

$$|\psi\rangle = a_1|\psi_1\rangle + a_2|\psi_2\rangle = \begin{pmatrix} \sqrt{n_{s1}}e^{i\varphi_1} \\ \sqrt{n_{s2}}e^{i\varphi_2} \end{pmatrix}. \tag{3.9}$$

Here the expansion coefficients are chosen so that they are normalized by the density of states, and the phase is detached: $a_{1,2} = \sqrt{n_{s1,2}}\exp(i\varphi_{1,2})$.

The system's Hamiltonian has the diagonal elements which are equal to the system's energy in the respective state: $H_{11} = \langle\psi_1|H|\psi_1\rangle = eV$ and $H_{22} = -eV$. The energy difference is equal to $2eV$, which corresponds to the Cooper-pair charge $2e$. The off-diagonal elements describe transitions between the levels, which are related to the tunneling through the barrier, which we characterize by the value B: $H_{12} = H_{21} = B$. Then the Hamiltonian becomes

$$H = \begin{pmatrix} eV & B \\ B & -eV \end{pmatrix} = eV\sigma_z + B\sigma_x. \tag{3.10}$$

And from the Schrödinger equation $i\hbar\frac{\partial}{\partial t}|\psi\rangle = H|\psi\rangle$ we obtain

$$\begin{aligned} i\hbar\dot{a}_1 &= eVa_1 + Ba_2, \\ i\hbar\dot{a}_2 &= Ba_1 - eVa_2 \end{aligned} \tag{3.11}$$

or, taking into account Eq. (3.9),

$$
\begin{cases}
\dot{n}_{s1} = -\dot{n}_{s2} = \dfrac{2Bn_{s0}}{\hbar} \sin \varphi, \\[2mm]
\dot{\varphi}_1 = -\dfrac{B}{\hbar} \cos \varphi - \dfrac{eV}{\hbar}, \\[2mm]
\dot{\varphi}_2 = -\dfrac{B}{\hbar} \cos \varphi + \dfrac{eV}{\hbar}.
\end{cases}
\tag{3.12}
$$

Here n_{s0} is the density of superconducting electrons in the banks without accounting for the weak link (this is slightly different from $n_{s1,2}$). Since the current through the tunneling junction is proportional to the rate of change of electron density, $I_s \propto dn_s/dt$, we get the formula for the *stationary Josephson effect*:

$$
I_J = I_c \sin \varphi,
\tag{3.13}
$$

where I_c is the junction critical current, which by assumption is much smaller than the critical current of the junction banks.

Subtracting the second equation of the system (3.12) from the third one, we obtain the formula for the *non-stationary Josephson effect*

$$
\dot{\varphi} = \frac{2eV}{\hbar}.
\tag{3.14}
$$

From this relation we get $\varphi(t) = \varphi^{(0)} + \frac{2e}{\hbar}Vt$. Then, for the supercurrent in (3.13) we obtain

$$
I_J = I_c \sin \left(\varphi^{(0)} + \frac{2e}{\hbar}Vt \right).
\tag{3.15}
$$

This means that applying the voltage results in flowing of the alternating current with the frequency

$$
\omega = \frac{2eV}{\hbar}.
\tag{3.16}
$$

Consequently, the average power, consumed from the external source for the supercurrent drive, is zero, $\overline{I_J V} = 0$. This means that the supercurrent, Eq. (3.15), does not dissipate energy. Furthermore, if we invert Eq. (3.14),

$$V = \frac{\hbar}{2e}\dot{\varphi}, \tag{3.17}$$

it becomes clear that the non-stationary Josephson effect consists in the appearance of the dc voltage on the junction, if the phase difference linearly depends on time.

So, if the junction is arranged with a stationary phase difference, $\phi = \phi^{(0)}$, and there is no voltage applied ($V = 0$), then the direct current flows through the junction $I_J = I_c \sin \varphi^{(0)} \leq I_c$. If the direct voltage is applied, then the alternating supercurrent appears, as described in Eq. (3.15). (At that, there is also a negligibly small normal current.)

Consider now the current-biased junction, i.e. the regime with the predetermined current (and not the voltage, as above) with value larger than the critical one $I > I_c$. It is important now to account for the normal current, which is defined by the value $I - I_c$. The full current is equal to the sum of the normal current V/R, with R being the resistance of the junction in normal state, and the supercurrent, Eq. (3.13),

$$I = I_c \sin \varphi + \frac{\hbar\dot{\varphi}}{2eR}. \tag{3.18}$$

Integrating this equation with respect to the time-dependent variable $\varphi(t)$, we can obtain, for the voltage $V = \frac{\hbar}{2e}\dot{\varphi}$, the following (see, for example, in Footnote 28 of page 85):

$$V(t) = R\frac{I^2 - I_c^2}{I + I_c \cos \omega t}, \quad \omega = \frac{2e}{\hbar}R\sqrt{I^2 - I_c^2}. \tag{3.19}$$

This means that, if the applied current is larger than the critical one, then the alternating voltage would appear on the junction. It can be shown that averaging over time gives

$$2e\bar{V} = \hbar\omega. \tag{3.20}$$

This can be interpreted as that the transfer of one Cooper pair through the junction is accompanied by the release of one quantum of electromagnetic radiation, which is known as *Josephson generation*.

From the definition of inductance as a proportionality coefficient between the current derivative and the voltage, $V = L(dI/dt)$, we obtain from Eqs. (3.13) and (3.14) the *Josephson inductance* of the junction:

$$L_J = \frac{V}{dI_J/dt} = \frac{\hbar}{2eI_c \cos \varphi}. \tag{3.21}$$

This means that a Josephson junction is described as a non-linear sign-changing inductance.

The relations above allow us to calculate also the *Josephson energy* of a junction:

$$E(\varphi) = \int I_J V dt = E_J (1 - \cos \varphi), \quad E_J = \frac{\hbar I_c}{2|e|}, \tag{3.22}$$

where we have defined characteristic energy of the junction E_J; at integration, the constant was chosen so as to have $E(\varphi = 0) = 0$.

These relations could also be obtained by considering the stationary Josephson effect as an equilibrium phenomenon. Then the current can be obtained by differentiating the thermodynamic potential: $I_J = c \, \partial F/\partial \Phi$, where the magnetic flux is related to the phase, $\Phi/\Phi_0 = \varphi/2\pi$. We obtain the expression for the respective Josephson energy: $F = c^{(-1)} \int I_J d\Phi = E_J (1 - \cos \varphi)$.

Besides the Josephson energy, E_J, a junction is also characterized by the electrostatic *charging energy* E_C. This energy is related to the charge $Q = C_J V$ on the plates of the tunnel junction with the capacitance C_J. The corresponding electrostatic (Coulomb) energy is $C_J V^2/2 = Q^2/2C_J$. Then the characteristic charging energy per electron is $E_C = e^2/2C_J$. (Note that in literature there is also another definition of the characteristic electrostatic energy, per Cooper pair, $(2e)^2/2C_J$.)

The ratio of the two characteristic energies, E_J and E_C, describes several types of qubits. The following three sections are devoted to the description of the basic types of superconducting qubits; see also Footnotes 16, 24 and 26 (on pages 51 and 84, respectively).

3.3. Current-biased junction

3.3.1. *Mechanical analogy*

This section is devoted to the simplest realization of a superconducting qubit, the *phase qubit*, which is based on a tunnel Josephson junction with current.

The scheme of physical realization is shown in Fig. 3.3(a), and the electric equivalent circuit is presented in Fig. 3.3(b). The junction is described by the resistance R connected in parallel with the capacitance C_J and the Josephson element J, in which flows the supercurrent $I_J = I_c \sin \varphi$.

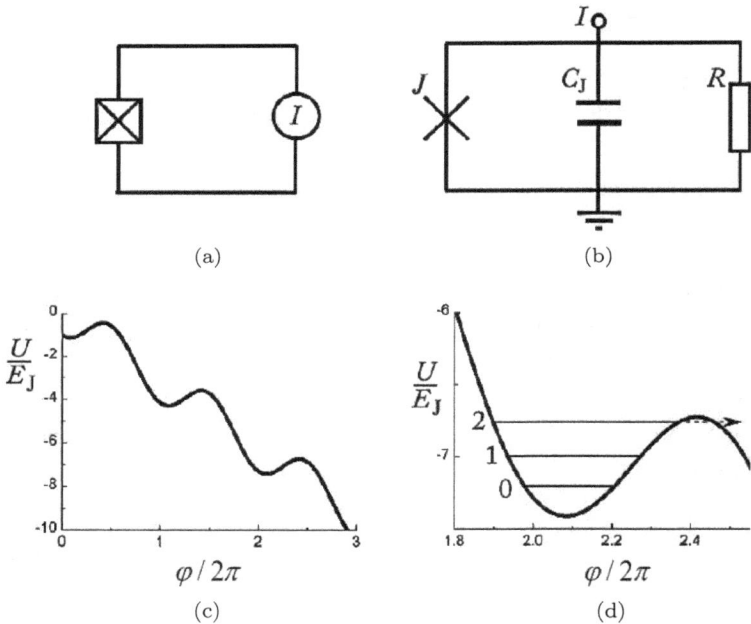

Fig. 3.3. (a) Scheme of a phase qubit, based on a current-biased Josephson junction. (b) Electric equivalent circuit. (c) Potential energy as a function of the order-parameter phase difference φ for $I_{dc}/I_c = 0.5$. (d) The same as in panel (c), but in the vicinity of the local minimum. The working levels of the qubit are shown and denoted by 0 and 1, as well as the next level 2, which is used for the state read-out.

For the dynamic variable φ — the order-parameter phase difference of the junction — one can write down the motion equation, using the Kirchhoff law for the full current as well as the formula (3.14), linking the voltage and the phase difference. This current is split into the normal component, V/R, the displacement current, $C_J dV/dt$, and the Josephson current in Eq. (3.13). We obtain

$$\frac{\hbar C_J}{2e}\frac{d^2\varphi}{dt^2} + \frac{\hbar}{2eR}\frac{d\varphi}{dt} + I_c\sin\varphi = I. \qquad (3.23)$$

The bias current can have both direct and alternating components, for example, in the form of a harmonic signal. In this section, for simplicity, we will not consider the alternating current. Then the motion equation (3.23) can be multiplied by $\hbar/2e$ and rewritten in the form of the equation for the mechanical motion of a particle with mass $m = (\Phi_0/2\pi)^2 C_J = \hbar^2/8E_C$, coordinate φ, experiencing a friction coefficient $\lambda = (\Phi_0/2\pi)^2 R^{-1}$ and potential energy $U(\varphi)$:

$$m\ddot{\varphi} + \lambda\dot{\varphi} = -\frac{dU}{d\varphi}, \qquad (3.24)$$

$$U(\varphi) = -E_J(\cos\varphi + \varphi I/I_c). \qquad (3.25)$$

The potential energy has the form of the washboard shown in Fig. 3.3(c). Such mechanical analogy allows description of classical dynamics of a Josephson junction. When the current is less than the critical value, $I < I_c$, there are local minima of the potential; the particle exhibits local oscillations and the average value of the phase difference is constant. According to Eq. (3.14), when that happens, the voltage at the junction is equal to zero. The small oscillations in the absence of a driving force (at $I = 0$) appear with the so-called plasma frequency of a Josephson junction (which will be defined later). At $I \geq I_c$ there are no local minima and the particle rolls down the board, which corresponds to the continuous change of the phase φ on the junction and the appearance of the voltage. So, at $I = I_c$ the plasma oscillations become unstable, and this corresponds to the transition to the dissipative regime, with the voltage across the junction.

3.3.2. *Quantization of the circuit*

Continuing the mechanical analogy, we can write down the Lagrangian and the Hamiltonian and quantize the system. For simplicity we neglect here the dissipation ($\lambda = 0$); the smallness of this value is necessary for practical applications. The electrostatic energy plays the role of the kinetic energy, $K = m\dot{\varphi}^2/2 = (\hbar^2/16E_C)\dot{\varphi}^2$, and the Josephson energy $U(\varphi)$ plays the role of the potential energy. Then we have for the Lagrangian: $L = K - U$. And the Lagrange equation,

$$\frac{d}{dt}\frac{\partial L}{\partial \dot{\varphi}} = \frac{\partial L}{\partial \varphi},$$

gives us the motion equation (3.24), or equivalent to this, Eq. (3.23).

Canonical momentum, conjugated to the canonical coordinate φ, is defined as $p = \partial L/\partial \dot{\varphi}$ and can be rewritten in the form of the charge on the junction, $Q = C_J V$, or in the form of the number of Cooper pairs on the junction, $n = Q/2e$, as follows:

$$p = m\dot{\varphi} = \frac{\hbar}{2e}C_J V = \hbar\frac{Q}{2e} = \hbar n, \qquad (3.26)$$

which defines the physical meaning of the generalized momentum.

The canonical Hamiltonian for the junction with current is

$$H(p, \varphi) = p\dot{\varphi} - L = \frac{p^2}{2m} + U, \qquad (3.27)$$

and it can be conveniently rewritten, using the Cooper-pair number n:

$$H(n, \varphi) = 4E_C n^2 - E_J(\cos\varphi + \varphi I/I_c). \qquad (3.28)$$

The quantization is done by the replacement in the Hamiltonian (3.27) of the canonical coordinate by the operator, $\varphi \to \hat{\varphi}$, and of the canonical momentum by the differential operator: $p \to \hat{p} = -i\hbar\partial/\partial\varphi$.[29] (In what follows, the hats of operators will be omitted

---

[29]Here a note about the phase operator is to the point. The problem of the operator definition relates to the 2π periodicity of the phase; then the minimal uncertainty of the Cooper-pair number n, at definite phase, would be $1/2\pi$ and not zero. Strictly speaking, the "phase operator" can only be introduced for large Cooper-pair number $n \gg 1$, which is the case here. See the detailed discussion in par. 2.2.1 in [Zagoskin 2011].

besides cases where this would lead to an ambiguity.) Instead of the momentum operator, it is more convenient to use the operator of the Cooper-pair number, $n = p/\hbar = -i\partial/\partial\varphi$, for which the commutation relation $[\varphi, p] = i\hbar$ gives $[\varphi, n] = i$. This commutation relation means that the phase and the number of Cooper pairs cannot be defined simultaneously:

$$\Delta n \Delta \varphi \geq 1. \tag{3.29}$$

They experience quantum fluctuations with two limiting cases. The first one takes place when the kinetic energy is dominating, at $E_C \gg E_J$; this describes the situation with small fluctuations of charge, $\Delta n \ll n$. Inversely, for $E_C \ll E_J$, the well-defined value is the phase. In the respective cases, the charge and phase (or flux, in the geometry of an interferometer) qubits are realized (see Footnote 16 on page 51).

3.3.3. *Phase qubit*

So, a quantum circuit, which consists of a current-biased Josephson junction, is described by the Hamiltonian (3.28). The quantization results in the appearance of discrete energy levels in the potential with local minima, as it is shown in Fig. 3.3(d). Since the potential differs from the harmonic one, the energy levels are non-equidistant. The lower two levels play the role of the operational qubit levels.

For the description of the phase qubit, we approximate the potential U by a parabola, expanding it near the minimum, where $U' = 0 = E_J(\sin\varphi - I/I_c)$ and $\varphi = \varphi_0 = \arcsin I/I_c$. Then, omitting the constant term, we have

$$U \approx E_J \cos\varphi_0 \frac{(\varphi - \varphi_0)^2}{2} \equiv m\omega_q^2 \frac{(\varphi - \varphi_0)^2}{2},$$

$$\omega_q^2 = \frac{E_J \cos\varphi_0}{m} = \frac{8E_J E_C}{\hbar^2}\sqrt{1 - \left(\frac{I}{I_c}\right)^2}. \tag{3.30}$$

Here we can also introduce

$$\omega_p = \sqrt{8E_J E_C}/\hbar, \tag{3.31}$$

which is the so-called *plasma frequency* — the frequency of small oscillations in the absence of the driving force ($I = 0$). The energy levels in such harmonic potential have the form $E_k = \hbar\omega_q(k + 1/2)$. So, the distance between the qubit energy levels, $\Delta E = E_1 - E_0 = \hbar\omega_q$, is defined by the bias current and by the plasma frequency of zero oscillations ω_p. We can now see that the phase qubit is the two-level approximation of a Josephson junction with current, when we can afford to disregard the upper levels. Such a qubit, similar to any other two-level system, can be described in terms of the Pauli matrices with the Hamiltonian $H_0 = -\frac{\Delta E}{2}\sigma_z$.

The principle of operation of the phase qubit is shown in Fig. 3.3(d). The bias current I is defined so that there are three energy levels in the well, the uppermost of which is close to the top of the potential barrier. The lower two levels are the working qubit levels, which are controlled by the excitations with the frequency equal to ω_q. These levels are far beneath the top of the potential-barrier, so the probability of tunneling is exponentially small. This corresponds to the localization in the potential well, that is $\bar{\varphi} = const$, and so the voltage across the junction is equal to zero, $V = \frac{\hbar}{2e}\dot{\bar{\varphi}} = 0$. In addition, the upper level is not excited, since the transition from level 1 to level 2 requires the frequency $\omega_{21} = (E_2 - E_1)/\hbar < \omega_q$. This level is involved in the read-out of the qubit state. Namely, the pulse is applied at the frequency ω_{21}; the transition to this level takes place and the probability of this process is equal to the occupation probability of level 1 — the upper qubit level — which requires measurement. Following which, tunneling from level 2 out of the potential well occurs and the voltage pulse $V = \frac{\hbar}{2e}\dot{\varphi}$ is registered. In this way, the probability of this voltage pulse is used to define the qubit population.

Note that introducing the characteristic frequency in Eq. (3.31) allows the inequalities for E_C and E_J, mentioned above, to be clarified. We have to compare respective terms of the Hamiltonian with the characteristic frequency, or, more precisely, with the energy gap between the levels, defined by this frequency, $\hbar\omega_p = \sqrt{8E_JE_C}$. Then, the almost classical regime, with the domination of the Josephson energy, is realized at $\hbar\omega_p \ll E_J$, that is at $E_C \ll E_J$.

3.4. Superconducting island — charge qubit

Consider the *charge superconducting qubit*. It is based on a super-conducting island, or the so-called *Cooper-pair box*.[30] The island is formed by a Josephson junction (with the energy E_J, capacitance C_J and phase difference φ) and the gate capacitance C_g, by means of which the island is connected to the gate electrode with the voltage V_g, as shown in Fig. 3.4(a). The smallness of the island and the capacitances provide large charging energy. We note that the island is in the regime called *Coulomb blockade*, where the Cooper pairs can tunnel discretely and the charge of the island is a well-defined value. In what follows, in Sec. 4.4, we will also discuss the Coulomb blockade for normal structures.

The electrostatic energy of the circuit, shown in Fig. 3.4(a), can be written for the island voltage $V = (\hbar/2e)\dot\varphi$ and transformed as follows:

$$\frac{C_J V^2}{2} + \frac{C_g(V_g - V)^2}{2} \rightarrow 4E_C \left(\frac{C_\Sigma V}{2e} - \frac{C_g V_g}{2e} \right)^2 \equiv 4E_C(n - n_g)^2.$$

$$(3.32)$$

Here we have defined the total capacitance of the island $C_\Sigma = C_J + C_g$ and the characteristic charging energy $E_C = e^2/2C_\Sigma$; during the

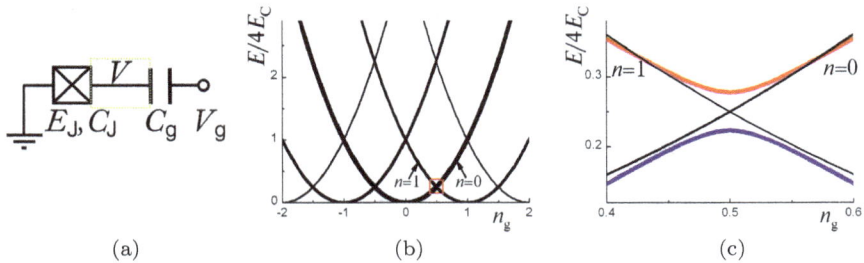

(a)   (b)   (c)

Fig. 3.4. (a) Schematic of the charge qubit. (b) Energy levels of the super-conducting island with the excess number of Cooper pairs $n = -2, -1, 0, 1, 2$. (c) Energy levels without accounting for the Josephson energy (two crossing lines) and the two-level approximation for $E_J/4E_C = 0.05$.

---

[30]M. Tinkham, *Introduction to Superconductivity*, 2nd ed., McGraw-Hill, New York (1996), Chapter 7.

transformation, we have omitted an insignificant constant. We have also defined the number of the Cooper pairs on the island, $n = \frac{C_\Sigma V}{2e} = \frac{C_\Sigma \hbar}{4e^2}\dot\varphi = \frac{\hbar}{8E_C}\dot\varphi$, and the dimensionless voltage on the gate electrode, $n_g = C_g V_g/2e$. Subtracting the junction Josephson energy, we obtain the system Lagrangian

$$L(\varphi, \dot\varphi) = 4E_C \left(\frac{\hbar}{8E_C}\dot\varphi - n_g\right)^2 + E_J \cos\varphi. \qquad (3.33)$$

Coming to the canonical momentum, $p = \partial L/\partial\dot\varphi = \hbar(n - n_g)$, we get the Hamiltonian

$$H(\varphi, p) = 4E_C(n - n_g)^2 - E_J \cos\varphi = \frac{4E_C}{\hbar^2}p^2 - E_J \cos\varphi. \qquad (3.34)$$

Next, we do the quantization as described in the previous section. We write down the Hamiltonian in the charge basis, that is the basis of the charge-operator eigenstates, $\hat{n}|n\rangle = n|n\rangle$. From here we have the expression for the charge operator, plus the completeness condition for the respective projectors:

$$\hat{n} = \sum_n n|n\rangle\langle n|, \quad \sum_n |n\rangle\langle n| = 1. \qquad (3.35)$$

Recall now (see [Landau and Lifshitz 1977], §15) that a particle wave function in the coordinate representation with the definite momentum $\mathbf{p}$ has the form $\psi_{\mathbf{p}}(\mathbf{r}) \equiv \langle\mathbf{r}|\mathbf{p}\rangle = e^{i\mathbf{r}\mathbf{p}/\hbar}$. In our case, the role of the generalized coordinate is played by φ and instead of the momentum, we use $n = p/\hbar$. Then we have the wave function $\langle\varphi|n\rangle = e^{in\varphi}$. (Also we have the commutation relation $[\varphi, n] = i$.) From this, using the completeness condition (3.35), we obtain the expansion

$$|\varphi\rangle = \sum_n |n\rangle\langle n|\varphi\rangle = \sum_n e^{-in\varphi}|n\rangle. \qquad (3.36)$$

Also, there is the inverse transformation (in all but name, the inverse Fourier transformation)

$$|n\rangle = \frac{1}{2\pi}\int d\varphi e^{in\varphi}|\varphi\rangle. \qquad (3.37)$$

As an exercise, let us check this:

$$|n\rangle = \frac{1}{2\pi} \int d\varphi e^{in\varphi} \sum_l e^{-il\varphi}|l\rangle = \sum_l |l\rangle \frac{1}{2\pi} \int_0^{2\pi} d\varphi e^{i(n-l)\varphi}$$

$$= \sum_l |l\rangle \delta_{nl} = |n\rangle. \tag{3.38}$$

Having in mind that we need the expression for $\cos \varphi$, it follows from Eq. (3.37) that

$$|n \pm 1\rangle = e^{\pm i\varphi}|n\rangle \quad \Rightarrow \quad (e^{i\varphi} + e^{-i\varphi})|n\rangle = |n+1\rangle + |n-1\rangle. \tag{3.39}$$

Here, in particular, we observed that the effect of the operator $\exp(il\varphi)$ is analogous to the finite-displacement operator $T_\mathbf{a} = \exp\left(\frac{i}{\hbar}\mathbf{a}\hat{\mathbf{p}}\right)$ such that $T_\mathbf{a}\psi(\mathbf{r}) = \psi(\mathbf{r}+\mathbf{a})$.

So, for the first and second terms in Eq. (3.34), we can write in the charge representation:

$$(n-n_g)^2 = \left(\sum_n n|n\rangle\langle n| - n_g\right)^2 = \left(\sum_n n|n\rangle\langle n|\right)^2$$

$$-2n_g \sum_n n|n\rangle\langle n| - n_g^2 \cdot \sum_n |n\rangle\langle n|$$

$$= \sum_n (n-n_g)^2|n\rangle\langle n|, \tag{3.40}$$

$$\cos\hat{\varphi} = \frac{1}{2}(e^{i\varphi} + e^{-i\varphi}) \cdot \sum_n |n\rangle\langle n|$$

$$= \frac{1}{2}\sum_n (|n+1\rangle\langle n| + |n-1\rangle\langle n|). \tag{3.41}$$

In the latter term, we make the change of the summation variable, and as a result, we obtain the Hamiltonian in the charge representation:

$$H = \sum_n \left\{ 4E_C(n-n_g)^2|n\rangle\langle n| - \frac{E_J}{2}(|n+1\rangle\langle n| + |n\rangle\langle n+1|) \right\}. \tag{3.42}$$

Here, the first, dominating, term describes the charging energy (recall that for the charge qubit, $E_C \gg E_J$). The respective energy levels $4E_C(n - n_g)^2$ are shown in Fig. 3.4(b). Here it is convenient to consider the excess Cooper-pair number on the island, rather than their total number. For this consideration, assume that the voltage is changed around some integer value $n_g \sim \bar{n}_g$, then $n - n_g = (n - \bar{n}_g) - (n_g - \bar{n}_g) \equiv N - N_g$, and in order to avoid introducing new variables, we change $N \to n$ and $N_g \to n_g$; then for illustration see Fig. 3.4.

In the two-level approximation, the Hamiltonian takes the form

$$H = 4E_C\{n_g^2|0\rangle\langle0| + (1 - n_g)^2|1\rangle\langle1|\} - E_J/2\{|0\rangle\langle1| + |1\rangle\langle0|\}. \tag{3.43}$$

We account for the completeness condition, $|0\rangle\langle0| + |1\rangle\langle1| = 1$, and obtain

$$2|1\rangle\langle1| = |1\rangle\langle1| + |1\rangle\langle1| + |0\rangle\langle0| - |0\rangle\langle0| = 1 + |1\rangle\langle1| - |0\rangle\langle0|. \tag{3.44}$$

Omitting the constant term (i.e. the one proportional to the unity matrix $|0\rangle\langle0| + |1\rangle\langle1|$), we get the expression for the Hamiltonian

$$H = -2E_C(1 - 2n_g)\{|0\rangle\langle0| - |1\rangle\langle1|\} - E_J/2\{|0\rangle\langle1| + |1\rangle\langle0|\}. \tag{3.45}$$

Or, introducing the Pauli matrices,

$$H = -\frac{\Delta}{2}\sigma_x - \frac{\varepsilon}{2}\sigma_z, \tag{3.46}$$

$$\Delta = E_J, \quad \varepsilon = 4E_C(1 - 2n_g). \tag{3.47}$$

So, the superconducting island can be described as a two-level system with controllable parameters. This is the *charge qubit*. The energy levels of the qubit can be obtained by diago- nalizing the Hamiltonian (3.46): $E_\pm = \pm(1/2)\sqrt{\Delta^2 + \varepsilon^2} = \pm 2E_C\sqrt{(1 - 2n_g)^2 + (E_J/4E_C)^2}$. These energy levels are plotted in Fig. 3.4(c).

In practice, they use the charge qubits with two junctions, embedded in a loop pierced by the external magnetic flux. This allows the Josephson energy to be made tunable. Consider for simplicity the case with two identical junctions with the critical currents I_c and the phase differences φ_1 and φ_2. Neglecting the loop geometric inductance, we have $\varphi_1 - \varphi_2 = 2\pi\Phi/\Phi_0$. The total current can then be written in the form

$$I = I_1 + I_2 = I_c(\sin\varphi_1 + \sin\varphi_2) = \tilde{I}_c \sin\varphi,$$

$$\tilde{I}_c = 2I_c \cos\pi\frac{\Phi}{\Phi_0}, \qquad \varphi = \frac{\varphi_1 + \varphi_2}{2}. \tag{3.48}$$

So, the loop with two junctions is described as a single junction with the critical current $\tilde{I}_c = \tilde{I}_c(\Phi)$, which corresponds to the effective Josephson energy $E_J(\Phi) = \hbar\tilde{I}_c/2e$. This allows the two parameters in the Hamiltonian (3.46) to be changed: $\varepsilon = \varepsilon(V_g)$ and $\Delta = \Delta(\Phi)$.

Consider separately the special type of charge qubits, the so-called *transmon*. Its schematic is shown in Fig. 3.5. This type of qubits probably harbours the most prospects to date,[31] due to the long decoherence times. The transmon itself consists of the charge qubit shunted by a large capacitance C_B and coupled to the *transm*ission-line resonator, which explains its name. The charge qubit in this case is in the form of the loop with two Josephson junctions, as was considered above. This loop is pierced by the external magnetic flux Φ from the current-carrying conductor situated nearby. This can be described as a single junction with the Josephson energy $E_J(\Phi) = E_{J0}\cos(\pi\Phi/\Phi_0)$. The qubit is coupled with the resonator, which is

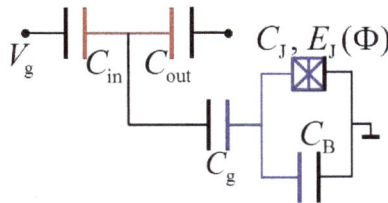

Fig. 3.5. Transmon: the charge qubit shunted by large capacitance C_B and coupled to the transmission-line resonator.

[31]G. Wendin, Quantum information processing with superconducting circuits: A review, Rep. Prog. Phys. **80**, 106001 (2017).

formed by the transmission line interrupted by two capacitances. Such a system, composed of a qubit and a resonator, presents much interest on its own for both fundamental science and applications, so we will devote a significant part of Chapter 5 to this.

The special feature of a transmon is its large shunting capacitance C_B, which results in a decrease in the charging energy of the superconducting island:

$$E_C = \frac{e^2}{2C_\Sigma} = \frac{e^2}{2\left(C_g + C_J + C_B\right)} \approx \frac{e^2}{2C_B}. \qquad (3.49)$$

This is done to decrease the curvature of the qubit energy levels, which is defined by the ratio $(E_J/E_C)^{-1}$, as is seen from the construction in Fig. 3.4(c). The shunting capacitance C_B decreases essentially this curvature. The ratio $E_J/E_C \sim 10$ is chosen; then the energy levels become almost flat and hence depend little on the charge fluctuations (noises). (True, this results in a decreased possibility of control by means of the gate voltage, so, now, the connection with the controlling electronics appears via the resonator.) Thanks to this, they manage to reach record decoherence times, of the order of tens of microseconds. For comparison, the time of a single operation on superconducting qubits is of the order of one nano-second. This means that before the phase is lost, one can make tens of thousands of operations, which is now enough to run quantum algorithms.

3.5. Ring with junctions

We have considered the phase qubit above. When such a qubit contains a loop (that is in the geometry of an interferometer), it becomes possible to control it by the external magnetic flux — such qubits are called *flux qubits*. The simplest realization would be a qubit in the form of a ring with one junction. And we will start from the discussion of such a classical circuit, having in mind its importance for applications. However, to create a double-well potential in such a circuit, a large inductance is required so that the respective energy would be of the order of the Josephson energy. But a large geometric inductance would require a large loop size and strong influence of the electromagnetic environment. In order to avoid this, the geometric inductance is replaced by the Josephson one. This allows the bistable situation to be achieved at a negligibly small geometric inductance of the loop. To date, the most popular circuit among researchers is the version with three junctions, while study the rings with larger number of junctions are also studied.

3.5.1. *Single-junction interferometer*

Consider a superconducting ring with a Josephson junction pierced by the magnetic flux Φ_e, see Fig. 3.1(b). This magnetic flux induces the circulating current I, and then the total magnetic flux in the ring with inductance L is

$$\Phi = \Phi_e - LI. \tag{3.50}$$

As we have seen, the total flux is defined by the phase difference on the junction, $\varphi = 2\pi\Phi/\Phi_0$. Accounting for the Josephson relation, we rewrite Eq. (3.50):

$$\varphi = \varphi_e - \beta \sin \varphi, \quad \varphi_e = 2\pi\Phi_e/\Phi_0. \tag{3.51}$$

Here the interferometer parameter was defined:

$$\beta = \frac{2\pi L I_c}{\Phi_0}. \tag{3.52}$$

Expressing the current from Eq. (3.50), $I = -(\Phi - \Phi_e)/L$, we obtain the following instead of Eq. (3.23)

$$\frac{\hbar C_J}{2e}\ddot{\varphi} + \frac{\hbar}{2eR}\dot{\varphi} + I_c \sin \varphi + \frac{\Phi_0}{2\pi L}(\varphi - \varphi_e) = 0. \tag{3.53}$$

Similar to the evolution of an autonomous Josephson junction considered above [see Eq. (3.24)], this corresponds to the mechanical motion in the potential

$$U = E_J(1 - \cos \varphi) + E_L\frac{(\varphi - \varphi_e)^2}{2}, \quad E_L = \frac{\Phi_0^2}{(2\pi)^2 L}. \tag{3.54}$$

Here E_L describes the magnetic energy; with this we can rewrite the interferometer parameter, $\beta = E_J/E_L$. The potential can be conveniently rewritten as

$$U = E_J\left(1 - \cos \varphi + \frac{(\varphi - \varphi_e)^2}{2\beta}\right). \tag{3.55}$$

If we plot the dependence $U(\phi)$, we would see that the potential has several minima only if $\beta > 1$. In particular, at $\varphi_e = \pi$ where the external flux is equal to half-integer flux quantum, the potential takes on the nature of a double-well. Incidentally, the minima correspond to the two opposite current directions in the ring.

However, the case of $\beta > 1$ corresponds to the large inductance L, for which a large ring is required, which would be strongly susceptible to external noise. Below, we will consider how to get the double-well potential at negligibly small inductance. But now we would like to make several remarks about one of the most important applications of the theory of superconductivity, which are the interferometers. For further reading, consult, for example, the textbook [Schmidt 1997].

The relation (3.51) defines implicitly the phase difference on a junction as a function of the external flux, $\varphi = \varphi(\varphi_e)$, and hence also the dependence of the current in the ring on the magnetic flux, $I = I(\Phi_e)$. (This dependence, in general, is multi-valued, with a hysteresis.) In practice, the current in such a ring is measured by inductively coupling it to a resonant circuit. Such a ring is called the *single-junction interferometer*, or *SQUID* — Superconducting QUantum Interference Device. Such devices are used for ultra-high-precision measurements of magnetic flux or other physical variables which can be converted to a magnetic flux. The modern SQUIDs can measure magnetic fields with a precision of up to 10^{-10} gauss (note, for comparison, that the Earth self-field is about 0.5 gauss) or a voltage with a precision of up to 10^{-15} V.

A single-junction SQUID is also called rf SQUID. In practice *two-junction interferometers*, or *dc SQUIDs*, are also used. Such a SQUID is a ring with two junctions, which is embedded in a circuit with direct current. This is similar to the schematic presented in Fig. 3.3, where one junction is replaced by a ring with two junctions. As we have seen, this results in the critical current depending on the magnetic flux, see Eq. (3.48).

3.5.2. *Flux qubit*

Consider a ring with three Josephson junctions, two of which have identical parameters, $E_{J1} = E_{J2} \equiv E_J$, and the third one is α

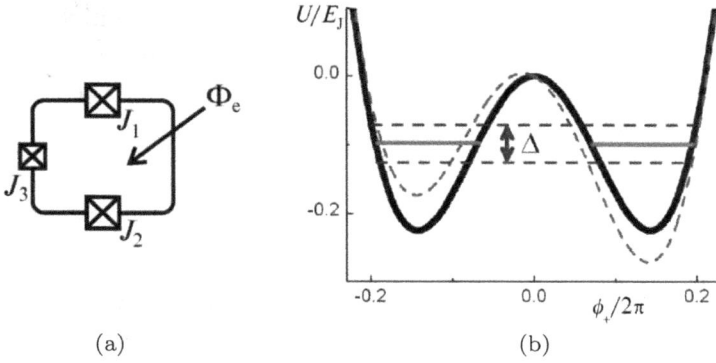

Fig. 3.6. (a) The flux qubit consists of a ring with three Josephson junctions, denoted as J_1, J_2, and J_3. The ring is pierced by a controlling external magnetic flux Φ_e. (b) Potential energy of the flux qubit with zero offset, $f = 0$, (thick line) and non-zero offset, $f = 0.01$ (dashed line); the parameter $\alpha = 0.8$. The horizontal lines show the energy levels in the wells in the cases of neglecting (solid lines) and correcting for tunneling effects (dashed lines).

times smaller, $E_{J3} = \alpha E_J$, see Fig. 3.6(a). The order-parameter phase difference on the three junctions equals φ_1, φ_2, and φ_3, respectively. Their sum is defined by the total flux in the ring, which approximately equals the external magnetic flux at very small geometric inductance:

$$\varphi_1 + \varphi_2 + \varphi_3 = 2\pi \frac{\Phi}{\Phi_0} = \frac{2\pi}{\Phi_0}(\Phi_e - LI) \approx 2\pi \frac{\Phi_e}{\Phi_0} \equiv 2\pi \left(\frac{1}{2} + f \right). \tag{3.56}$$

The potential energy of the ring is defined by the Josephson energies of the junctions:

$$U = -E_J(\cos \varphi_1 + \cos \varphi_2 + \alpha \cos \varphi_3)$$
$$= -E_J(2 \cos \varphi_- \cos \varphi_+ - \alpha \cos(2\pi f - 2\varphi_+)), \tag{3.57}$$

where the phase φ_3 was excluded using the relation (3.56) and we also defined $\varphi_\pm = (\varphi_1 \pm \varphi_2)/2$. The potential profile has two local minima, which at $f = 0$ are defined by the values $\varphi_- = 0$ and $\varphi_+ = \pm\varphi_0 = \pm \arccos(1/2\alpha)$ $(U'/E_J = 2 \sin \varphi_+ - 2\alpha \sin 2\varphi_+)$. Classical motion between these two minima is restricted by the high potential barriers in the direction of changing φ_-. Therefore we can consider the motion as being one-dimensional at $\varphi_- = 0$ with the potential energy

$$U(\varphi_+, f) = -E_J(2\cos\varphi_+ - \alpha\cos(2\pi f - 2\varphi_+)), \qquad (3.58)$$

which depends on the generalized coordinate φ_+ and the external magnetic flux f. The potential becomes the double-well potential for the values f close to 0 and α between 0.5 and 1. This is demonstrated in Fig. 3.6(b).

The local minima, where $\partial U/\partial\varphi_+ = 0$ at $\varphi_+ = \pm\varphi_0$, define the states with the persistent current in the ring with opposite directions and the amplitudes

$$I_p = I_c\sin(\varphi_+ + \varphi_-) = \pm I_c\sin\varphi_0 = \pm I_c\sqrt{1 - 1/(2\alpha)^2}. \qquad (3.59)$$

These current-carrying states define the basis of the flux qubit. The energy levels for these states are shown by the solid lines in Fig. 3.6(b). A non-zero offset f moves apart these levels by the value $\delta U = 2I_p\Phi_0 f \equiv \varepsilon$. And the quantum-mechanical tunneling with the amplitude Δ results in the lifting of the degeneracy in energy and splitting of the levels.

So, we come to the fact that the flux qubit is described by the two-level Hamiltonian, which formally coincides with that for a charge qubit, Eq. (3.46), with the difference that the values Δ and ε now have different meaning. Also, the Hamiltonians are defined in different bases — the charge and the current bases for these two types of qubits. In practice, the qubits often have constant value Δ, while the bias ε contains both constant component ε_0 and the alternating one $\tilde{\varepsilon}(t)$. The former bias parameter, ε_0, changes the distance between the energy levels $\Delta E = \sqrt{\Delta^2 + \varepsilon_0^2}$, and the latter driving parameter, $\tilde{\varepsilon}(t)$, controls the population of these levels. For most of the applications, it is convenient to consider the alternating part of the bias in the form of a harmonic signal.

Conclusion to Chapter 3

In summary, let us write down the Hamiltonian, which describes qubits, in the form

$$H(t) = -\frac{\Delta}{2}\sigma_x - \frac{\varepsilon(t)}{2}\sigma_z, \tag{3.60}$$

$$\varepsilon(t) = \varepsilon_0 + A\sin\omega t. \tag{3.61}$$

As a rule in studying dynamics, we consider the value Δ constant and the other parameters — ε_0, A and, ω — to be the control parameters. They can be changed by means of either the external current, gate voltage, or magnetic flux — in dependence on the specific type of the circuit (qubit) — for a phase, charge, or flux qubits, respectively. The Hamiltonian (3.60) is written in the computational basis which corresponds to diverse observable quantities for different types of qubits. In particular, for a flux qubit, these are the states with the current in clockwise and counterclockwise directions; for a charge qubit, these are the number of the excess Cooper pairs on the island. Corresponding to the foundations of quantum mechanics, one would expect the possibility of finding such systems in the superposition of these macroscopic states. This was reliably confirmed by numerous experiments.

Let us now present, for the readers' information, basic quantitative characteristics of superconducting qubits:

* Distance between energy levels $\Delta/h \sim 1 - 10\,\text{GHz}$ (which corresponds to excitation and emission by microwave photons);
* thickness of a Josephson junction $\sim 1\,\text{nm}$, and the scale for a circuit $\sim 1\mu\text{m}$;
* effective temperature $T \sim 50\,\text{mK}$ (which corresponds to 1 GHz h/k_B; it is necessary to have $k_B T < \Delta$);
* characteristic times for relaxation and decoherence $T_{1,2} \lesssim 10\,\mu\text{s}$;
* ratio E_J/E_C has different orders of magnitude as listed: for charge qubits — 0.1, for flux qubits — 10, and for phase qubits — 10^6.

Problems for independent work and for self-assessment

3.1. (***) Based on the two-level model, describe the two Josephson effects.

3.2. (**) Solve the differential equation for the current, Eq. (3.18).

3.3. (**) Average, over one period, the voltage on the junction, in Eq. (3.19), which would lead to the relation for the Josephson generation.

3.4. (**) Based on the Kirchhoff law, write down the equation of motion for a current-biased Josephson junction and draw its potential energy.

3.5. (**) Expanding in series the potential energy from the previous task, get the characteristic frequency of the phase qubit.

3.6. (**) Analyzing the electrostatic energy of a superconducting island, write down its Hamiltonian.

3.7. (***) Quantize the Hamiltonian of the island and write it in the charge basis.

3.8. (*) Write down the island Hamiltonian in the two-state approximation and obtain the pseudo-spin Hamiltonian of the charge qubit.

3.9. (**) Obtain and sketch the potential energy of a single-junction interferometer.

3.10. (***) Plot the potential energy of the three-junction ring as a function of φ_- and φ_+, and be convinced that it has steep walls in the direction of increasing φ_-, which allows us to assume $\varphi_- = 0$.

3.11. (**) Demonstrate that the three-junction ring (after the previous task) can be considered as a flux qubit; for this, sketch the potential energy as a function of φ_+ and analyze it.

Chapter 4

NORMAL QUANTUM CIRCUITS

> "When studying science, the examples are more useful than the rules."
>
> [I. Newton]

Many of the quantum properties, inherent to superconducting circuits, can be realized in normal-metal coherent systems. This relates to systems of size much smaller than the coherence length L_φ. The phase coherence means that the particles can be described by a wave function, which evolves according to the Schrödinger equation. In the present chapter we will describe such systems, paying particular attention to specific properties such as conductance quantization and Coulomb blockade, which are important for quantum engineering. For further reading we recommend the textbooks [Moskalets 2010], [Zagoskin 2011] and the references in Footnotes 32 and 33 below.

4.1. Low-dimensional structures

Such systems are realized on the basis of a *two-dimensional electron gas* (2DEG), in which electrons are restricted from moving along one dimension, with their freedom of mobility confined to a plane, perpendicular to that dimension. Such a situation appears in a

---

[32]S. Datta, *Electronic Transport in Mesoscopic Systems*, Cambridge, UK: Cambridge University Press (1997).

[33]E. Akkermans and G. Montambaux, *Mesoscopic Physics of Electrons and Photons*, Cambridge, UK: Cambridge University Press (2007).

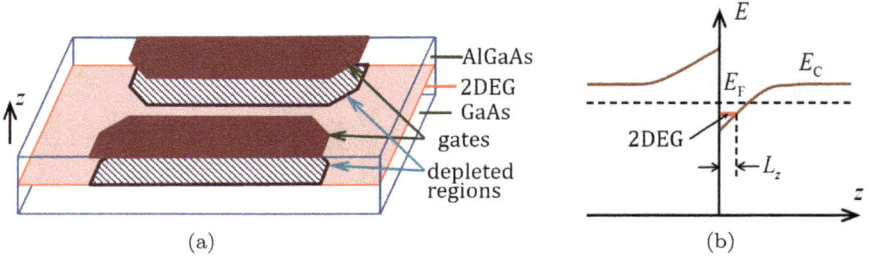

Fig. 4.1. (a) Formation of a two-dimensional electron gas (2DEG) on the heterojunction GaAs/Al$_x$Ga$_{1-x}$As. Application of the voltage to gates creates areas of geometrical shadow, which are not accessible to the electrons of the 2DEG. At a certain configuration of the gates, they create quantum wires and dots, which will be described later. (b) Potential energy around the boundary. Asymmetric potential results in quantization in the well with a single energy level for electrons forming 2DEG.

heterojunction of two semiconductors, for example, gallium arsenide and aluminium-doped gallium arsenide, GaAs/Al$_x$Ga$_{1-x}$As, which is schematically shown in Fig. 4.1(a). In such a heterojunction, due to the incommensurability of the lattices, the curvature of the conduction bands appears. As a result, in the vicinity of the junction, the potential well arises, as it is shown in Fig. 4.1(b). It is important, first, that the quantization in the transverse direction results in there being only one energy level in the potential well; second, the electrons possess high mobility in the longitudinal direction; lastly, the collective electrons can be described as quasiparticles with the electron charge e and mass m, which is different, generally speaking, from the free electron mass m_e. Additional electrodes (or the gates, as they say) allow "press out" of electrons from the area of their geometrical shadows by means of the negative gate voltage, $V_g < 0$, as shown in Fig. 4.1(a). In this way, the low-dimensional structures are formed. The characteristic parameters for a 2DEG are the following: coherence length $L_\varphi \sim 1\,\mu$m, Fermi wavelength $\lambda_F \sim 50$ nm (note the difference from conventional metals, where $\lambda_F \sim 0.1$ nm), effective mass of the quasiparticles $m \simeq 0.07 m_e$ ($m_e \approx 10^{-30}$ kg), density $n \sim 10^{12}$ cm^{-2}.

To describe low-dimensional structures, consider the *spectrum quantization* for them in the simplest model, when electrons are confined in a three-dimensional well of the size $L_x \times L_y \times L_z$. Then, the assumption of the phase coherence allows the system to be described in a single-particle approximation with the Schrödinger equation for conduction electrons (quasiparticles) with the effective mass m:

$$-\frac{\hbar^2}{2m} \sum_{i=1}^{3} \frac{\partial^2}{\partial x_i^2} \psi = E\psi. \tag{4.1}$$

Here $x_1 = x$, $x_2 = y$, $x_3 = z$. We assume that our three-dimensional potential well has infinitely high walls, which means that we apply the zero boundary conditions on the wave function at $x_i = 0, L_i$. We obtain the solution by separating the variables and normalizing the wave function to unity:

$$\psi(x, y, z) = \psi_1(x)\psi_2(y)\psi_3(z),$$

$$\psi_i(x_i) = \sqrt{\frac{2}{L_i}} \sin k_i x_i,$$

$$k_i = \frac{\pi n_i}{L_i}, \quad n_i = 1, 2, \dots, \tag{4.2}$$

$$E = \sum_{i=1}^{3} E_i, \quad E_i = \frac{\hbar^2 k_i^2}{2m} = \frac{\hbar^2 \pi^2}{2m} \frac{n_i^2}{L_i^2}.$$

Thus, the state of the electrons in the three-dimensional potential well is defined by the set of three numbers n_i. The respective basis functions and energies are the following: $\psi = \prod_{i=1}^{3} \psi_i^{(n_i)}(x_i) = \psi_1^{(n_1)}(x)\psi_2^{(n_2)}(y)\psi_3^{(n_3)}(z)$ and $E_{n_1,n_2,n_3} = \sum_i E_i^{(n_i)}$.

The maximally filled number of levels $n_i^{\max} \equiv N_i$ is defined by the chemical potential μ, which in the absence of fields equals the Fermi energy ε_F. Then the maximal wave number equals $k_F = p_F/\hbar = \sqrt{2m\varepsilon_F}/\hbar$, and for the maximal number of the levels we get

$$N_i = \left[\frac{L_i k_F}{\pi}\right] = \left[\frac{L_i}{\lambda_F/2}\right], \tag{4.3}$$

where the Fermi wavelength is $\lambda_F = 2\pi/k_F$. This result means, in particular, that, if $L_i < \lambda_F/2$, none of the energy levels is filled and the current cannot flow through such samples at low voltages.

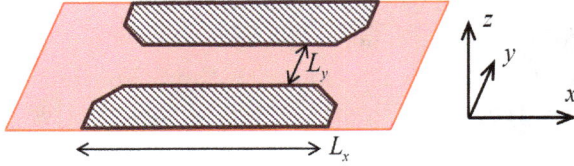

Fig. 4.2. Quantum wire: in the plane of the 2DEG where the regions inaccessible for electrons are shaded; the wire length is $L_x \gg \lambda_F$ and the width is $L_y \geq \lambda_F/2$.

For the transverse quantization in heterojunctions, which we discussed above, the potential is such that its characteristic width L_z is larger than or of the order of $\lambda_F/2$. And so the conductivity of the 2DEG is defined by a single subband with $n_z = 1$; the energy of 2DEG is then counted from $E_z = \hbar^2\pi^2/2mL_z^2$. If other dimensions are also of the order of the Fermi wavelength, as in Fig. 4.2, then such a system becomes one- or zero-dimensional. Such conductors are called *quantum wires* and *quantum dots*, respectively. Note that if the system is large in size along the x direction, $L_x \gg \lambda_F$, then $N_x \gg 1$ and the distance between the energy levels becomes small, $\Delta E_x \sim \varepsilon_F/N_x \rightarrow 0$, therefore the spectrum can be considered continuous: $E_x = p^2/2m$.

We note separately that the potential, which forms 2DEG, is essentially asymmetric. This corresponds to the transverse electric field $\vec{E} = E_0\vec{e}_z$. Hence, for such structures in some problems, it is essential to take into account the so-called *Rashba spin-orbit interaction*. In the general case, the spin-orbit interaction appears in relativistic quantum mechanics by expanding the Dirac equation in the powers of $1/c$. Another illustrative way of deriving the spin-orbit Hamiltonian is by making use of the Lorentz transformations of the electromagnetic field. Next, the electric field $\vec{E}$ in the coordinate system, bound with the electron, acts as the effective magnetic field $\vec{B} = \frac{1}{c^2}\vec{E} \times \vec{v} = \frac{1}{mc^2}\vec{E} \times \vec{p}$. Then the spin-orbit Hamiltonian is given by the Zeeman interaction of the electron spin with this field:

$$H_{SO} = \frac{g}{2}\mu_B\vec{\sigma}\vec{B} = \frac{g\mu_B E_0}{2mc^2}\vec{\sigma} \cdot \vec{e}_z \times \vec{p} \equiv \alpha\vec{p} \times \vec{\sigma} \cdot \vec{e}_z. \qquad (4.4)$$

The value α introduced here appears to be sufficiently large: even though it is inversely proportional to the squared light velocity, the potential energy demonstrated in Fig 4.1(b) changes over atomic-size distance, and therefore the electric field E_0 has a significant value. On the other hand, the respective energy should be compared with the Fermi energy, and it is small for 2DEG — much less than for the conduction electrons in 3D conductors. Thereby, a series of problems for low-dimensional structures is solved with addition of the term in Eq. (4.4) to the Hamiltonian of free conduction electrons. Having introduced this spin-orbit interaction here, in our first-iteration approach to this field, we will neglect this term in what follows.

4.2. Conductance quantization

Consider the conductivity of a quantum wire — a one-dimensional conductor — which is connected to bulk conductors as shown in Fig. 4.3(a). These conductors play the role of the reservoirs, which are characterized by certain values of the temperature T and the chemical potential μ, and accordingly the electrons in them are described by the Fermi distribution function. We assume that the voltage is applied to the banks of the contact, $V > 0$, and respectively the chemical potentials are shifted by the value of the electric potential, $\pm eV/2$. The wire is assumed to be *ballistic*, which means that the dephasing length is larger than its length, $L_\varphi \gg L$. We

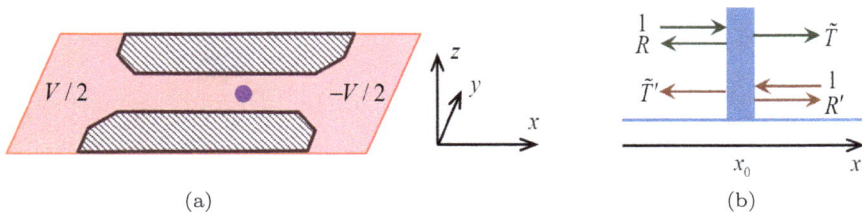

Fig. 4.3. (a) Quantum wire with a single scatterer. By means of the additional gates (not shown in the schematic), the voltage V is applied. (b) Transmission and reflection coefficients for the electronic waves, incident from the left (shown at the top) and from the right (shown at the bottom).

further assume that there is a scatterer (an impurity) within the wire, which is characterized by the reflection coefficients.

The solution of the Schrödinger equation, $i\hbar\frac{\partial\Psi}{\partial t} = -\frac{\hbar^2}{2m}\frac{\partial^2}{\partial x^2}\Psi$, is the wave function

$$\Psi_p(x,t) = \frac{1}{\sqrt{L}}e^{-i\frac{\varepsilon}{\hbar}t}e^{i\frac{p}{\hbar}x}, \quad \varepsilon = \frac{p^2}{2m}. \tag{4.5}$$

m, p, and ε are respectively the mass, momentum, and kinetic energy of an electron, and $L \equiv L_x$. The wave function of an electron is normalized to unity, $\int_0^L dx\,|\Psi|^2 = 1$. The current, created by the electron motion with momentum p, can be calculated as follows:

$$I_p = e\frac{i\hbar}{2m}\left(\Psi\frac{\partial}{\partial x}\Psi^* - \Psi^*\frac{\partial}{\partial x}\Psi\right) = \frac{e\hbar}{m}\mathrm{Im}\left(\Psi^*\frac{\partial}{\partial x}\Psi\right) = \frac{ev}{L}, \tag{4.6}$$

where $v = p/m$ is the electron velocity.

To obtain the total current, we have to sum over all the states, accounting for their occupations by multiplying to the distribution function $f(p)$. Assuming that the conductor is long ($L \gg \lambda_F$), we can change the summation to an integration. Indeed, for $p = p_n = \hbar\frac{\pi n}{L}$, we have $\Delta p = p_{n+1} - p_n = \pi\hbar/L \to 0$. Therefore we can use the relation $1 = \frac{L}{\hbar\pi}\Delta p$ and make the replacement $\Delta p \to dp$ to obtain

$$\sum_n \cdots = \frac{L}{\pi\hbar}\sum_n \Delta p \cdots \to \frac{L}{\pi\hbar}\int_0^{p_F} dp \cdots \approx \frac{L}{2\pi\hbar}\int_{-\infty}^{\infty} dp \cdots \tag{4.7}$$

Then for the total current, accounting for the two-fold degeneracy in spin, we have

$$I = \sum I_p f(p) \approx 2\frac{L}{h}\int_{-\infty}^{\infty} dp I_p f(p) = \frac{2e}{mh}\int_{-\infty}^{\infty} dp\, p f(p)$$

$$= \frac{2e}{mh}\int_0^{\infty} dp\, p\,(f(p) - f(-p)). \tag{4.8}$$

Let us compute for the integration with respect to the energy $\varepsilon = p^2/2m$; then $d\varepsilon = pdp/m$ and we obtain

$$I = \frac{2e}{h} \int_0^\infty d\varepsilon \, (f_>(\varepsilon) - f_<(\varepsilon)), \qquad (4.9)$$

where $f_>(\varepsilon)$ and $f_<(\varepsilon)$ correspond to electrons with $v > 0$ and $v < 0$, respectively. Let us discuss these distribution functions. Electrons incident from the left and right banks are described by the Fermi distribution functions

$$f_{1,2} = f_F(\varepsilon - \mu_{1,2}) = \left[1 + \exp\left(\frac{\varepsilon - \mu_{1,2}}{k_B T}\right)\right]^{-1} \qquad (4.10)$$

with respective chemical potentials $\mu_{1,2} = \mu \pm eV/2$. The electrons from the left bank with the probability $\tilde{T}$ pass by the scatterer and those with the probability R are reflected from it; the electrons from the right bank with the probability $\tilde{T}'$ are transmitted and those with the probability R' are reflected; see Fig. 4.3(b). The probabilities of the states' occupations are characterized by the distribution functions; to the right from the scatterer we have: $f_<(\varepsilon) = f_2$ and $f_>(\varepsilon) = \tilde{T}f_1 + R'f_2$. Then from Eq. (4.9), accounting for the fact that $1 - R' = \tilde{T}'$, we obtain

$$I = \frac{2e}{h} \int_0^\infty d\varepsilon \left(\tilde{T}f_1 - \tilde{T}'f_2\right). \qquad (4.11)$$

Next, we assume that the potential barrier, associated with the scatterer, is symmetric so $\tilde{T}' = \tilde{T}$. Consider, in what follows, the response linear in the voltage, which means that we assume $|eV| \ll \mu$. Then we can expand the distribution functions (4.10) into series:

$$f_F\left(\varepsilon - \mu \mp \frac{eV}{2}\right) \approx f_F(\varepsilon - \mu) \mp \frac{eV}{2} f_F'(\varepsilon - \mu), \qquad (4.12)$$

$$f_F' = -\frac{1}{4k_B T} \cosh^{-2} \frac{\varepsilon - \mu}{2k_B T} \xrightarrow[T \to 0]{} -\delta(\varepsilon - \mu). \qquad (4.13)$$

Then, for the *conductance* of the quantum wire, we obtain the so-called *Landauer formula*:

$$G = \frac{I}{V} = G_0 \int_0^\infty d\varepsilon \tilde{T}(\varepsilon) \left(-\frac{\partial f_F}{\partial \varepsilon} \right), \qquad (4.14)$$

where we have introduced the *conductance quantum*

$$G_0 = \frac{2e^2}{h}. \qquad (4.15)$$

At zero temperature (i.e. at $T \ll \mu$), and accounting for Eq. (4.13), we obtain for the conductance:

$$G|_{T=0} = G_0 \tilde{T}(\mu). \qquad (4.16)$$

This formula reduces the problem of kinetics — which is the definition of the current — to the quantum-mechanical problem of scattering.

We emphasize that even in the absence of a scatterer, when the transmission probability $\tilde{T} = 1$, the conductance of a quantum wire is not infinite — such a one-dimensional conductor creates the resistance $R_0 = 1/G_0 \approx 13$ kΩ. Note that $G_0 \approx 7.8 \cdot 10^{-5}$ siemens (1 siemens = 1 Ω^{-1}), and one can find in literature another definition of a characteristic conductance without the factor "2" in Eq. (4.15).

Finally, let us discuss the situation where the quantum wire is not a strictly one-dimensional conductor, but rather a conductor of finite width L_y — a two-dimensional conducting channel. Then the number of the levels which are filled is $N_y = [2L_y/\lambda_F]$, corresponding to the transverse quantization. We have to sum over them, and then the Landauer formula takes the form [Moskalets 2010, Chapter 8]: $G|_{T=0} = G_0 \sum_{n=1}^{N_y} \tilde{T}_n(\mu)$. Note that the number N_y depends on the channel width L_y, which in an experiment is defined by the gate voltage V_g. So, the Landauer formula describes the stepwise dependence of the conductance on the voltage. If $\tilde{T}_n = 1$, then as illustrated in Fig. 4.4(b), the height of the steps equals G_0 in the dependence of the conductance G on the channel width L_y. If $\tilde{T}_n < 1$, then the conductance decreases correspondingly, and the resistance of the conductor increases. We emphasize that for a

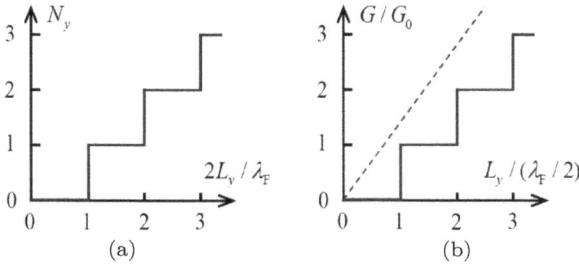

Fig. 4.4.   Dependence of the number of conducting subbands N_y (a) and of the conductance G (b) on the conductor width L_y. The height of the conductance steps is G_0, and their width is $\lambda_F/2$.

macroscopic conductor, the conductivity is $G = \sigma \frac{S}{L}$, which is linearly dependent on the conductor width, while for a mesoscopic conductor this dependence is stepwise.

4.3. Aharonov–Bohm effect

Mesoscopic systems allow the study of fundamental effects such as the Aharonov–Bohm effect in quantum mechanics, where the scalar and vector electromagnetic potentials directly (via the phase of a particle wave function) define the observable values. This means that they have physical meaning on their own. In contrast, observables in classical physics are defined by electric and magnetic fields only, while the potentials are introduced only as a convenient parametrization. And so, for the electric field and the magnetic-field induction we have: $\mathbf{B} = \mathrm{rot}\,\mathbf{A}$ and $\mathbf{E} = -\nabla\varphi - \frac{1}{c}\frac{\partial\mathbf{A}}{\partial t}$ (this is written in the CGS system, while in SI: $\mathbf{B} = \mathrm{rot}\,\mathbf{A}$ and $\mathbf{E} = -\nabla\varphi - \frac{\partial\mathbf{A}}{\partial t}$). Such "parametrization" corresponds to the Maxwell equations in vacuum: $\mathrm{rot}\,\mathbf{E} = -\frac{1}{c}\frac{\partial\mathbf{B}}{\partial t}$ and $\mathrm{div}\,\mathbf{B} = 0$. Note that the Maxwell equations are not changed by the gauge transformation: $\mathbf{A} \to \mathbf{A} + \nabla\chi$ and $\varphi \to \varphi - \frac{1}{c}\frac{\partial\chi}{\partial t}$.

4.3.1. *Conductance oscillations*

Consider the manifestation of the Aharonov–Bohm effect in a mesoscopic sample. For this, let us calculate the transmission of the electron current through a doubly-connected sample — a ballistic

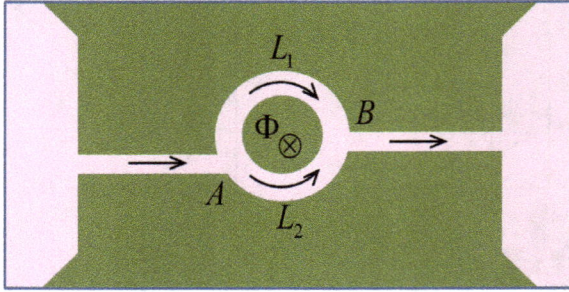

Fig. 4.5. Propagation of the current through the doubly-connected mesoscopic sample pierced by the magnetic flux Φ.

ring. Let this ring be pierced by the magnetic flux Φ. This could be a weak magnetic field — so weak that its influence (via the Zeeman term) can be neglected and only changes of the phase factor should be taken into account. But more illustrative would be to imagine a thin solenoid inside the ring. The vector potential of such a field is directed tangentially to the circle around the solenoid and depends only on the circle radius R, that is on the distance from the solenoid: $A = \Phi/2\pi R$. The magnetic-field induction outside the solenoid is $\mathbf{B} = \mathrm{rot}\,\mathbf{A} = 0$. Therefore in such a formulation of the problem in classical physics, the presence of the solenoid field does not influence the current in the circuit, as in Fig. 4.5.

For the mesoscopic formulation of the problem, we need to discuss the propagation of the electronic wave — the splitting in point A and the interference in point B. As we have seen above, the conductance is defined by the transmission coefficient.

The wave function of a free electron, moving along a certain trajectory, is defined by the Schrödinger equation $(1/2m)(-i\hbar\nabla - e\mathbf{A}/c)^2\psi = E\psi$ and has the form:

$$\psi(\mathbf{r}') = const \cdot \exp\left[\frac{i}{\hbar}\int_{\mathbf{r}_0}^{\mathbf{r}'} d\mathbf{r}\,(\mathbf{p} - e\mathbf{A}/c)\right] = \psi(\mathbf{r}_0) \cdot a(\mathbf{r}', \mathbf{r}_0).$$

(4.17)

Here we have fixed the lower integration limit at a certain point $\mathbf{r}_0$, therefore the normalizing constant is defined as $\psi(\mathbf{r}_0)$. The value $a(\mathbf{r}', \mathbf{r}_0)$ is the probability amplitude of the transmission from one point to another. Correspondingly, for the propagation of the electronic wave along the upper and lower parts of the ring, we have

$$a(\mathbf{r}_{B1,2}, \mathbf{r}_A) \equiv a_{1,2} = |a_{1,2}| e^{i\delta_{1,2}}, \quad \delta_{1,2} = \delta_{1,2}^{(0)} - \frac{e}{\hbar c} \int_{1,2} \mathbf{A} d\mathbf{l}.$$

$$(4.18)$$

Here we have separated the term $\delta_{1,2}^{(0)}$ independent of the vector potential. In the absence of external forces, the momentum is conserved hence $\delta_{1,2}^{(0)} = \frac{p}{\hbar} L_{1,2}$. Then for the total probability of the electron transmission by adding the amplitudes,

$$|a_1 + a_2|^2 = |a_1|^2 + |a_2|^2 + 2 |a_1 a_2| \cos(\delta_1 - \delta_2). \qquad (4.19)$$

And for the phase difference, corresponding to propagations along the upper and lower branches, we get

$$\delta_1 - \delta_2 = \delta_1^{(0)} - \delta_2^{(0)} - \frac{e}{\hbar c} \oint \mathbf{A} d\mathbf{l} = \Delta \delta^{(0)} + \frac{2\pi \Phi}{\Phi_0^N}, \qquad (4.20)$$

where we have defined the *magnetic flux quantum*,

$$\Phi_0^N = \frac{hc}{|e|}, \qquad (4.21)$$

which, distinct from the case of superconductors, contains the electron charge e, rather than the charge of a Cooper pair $2e$. ($\Phi_0^{(N)} = \frac{hc}{|e|} \approx 4 \cdot 10^{-7}$ gauss $\cdot$ sm^2 in the system of units CGS and $\Phi_0^{(N)} = \frac{h}{|e|} \approx 4 \cdot 10^{-15}$ Wb in SI.) In what follows, we will omit the index N from Φ_0^N.

Taking into account of that, the conductance is proportional to the electron transmission coefficient, satisfying Landauer formula, and we may obtain

$$G = G_1 + G_2 + 2\sqrt{G_1 G_2} \cos\left(\Delta \delta^{(0)} + \frac{2\pi \Phi}{\Phi_0}\right), \qquad (4.22)$$

where $G_{1,2}$ are the conductances of the upper and lower parts of the circuit. In a symmetric formulation of the problem, $G_1 = G_2$, and the conductance is a periodic function of the magnetic flux:

$$G = 2G_1 \left[1 + \cos \left(\Delta\delta^{(0)} + \frac{2\pi\Phi}{\Phi_0} \right) \right] \in (0 \cdots 4G_1), \qquad (4.23)$$

where the minimal value, 0, corresponds to the destructive interference of the two waves and the maximal value, $4G_1$, is the result of the constructive interference. This is one more radical distinction from the case of classical conductors, where there is no interference and the conductances sum up: $G = 2G_1$.

   * We note that the full solution of the problem on the transmission through our doubly-connected conductor, Fig. 4.5, (which assumes the solution of the Schrödinger equation in each of the branches and then the matching of the respective wave functions) would give the following expression for the transmission coefficient of an electron with the wave number $k = p/\hbar$ [Moskalets 2010]:

$$T(k) = \frac{4\sin^2(kL/2)\cos^2(\pi\Phi/\Phi_0)}{[1 + \cos(2\pi\Phi/\Phi_0) - 2\cos kL]^2 + \sin^2 kL}, \qquad (4.24)$$

where L is the ring circumference, and correspondingly $G = G_0 T(k_F)$.

4.3.2. *Persistent current*

Consider one more important visualization of the Aharonov–Bohm effect, which is the possibility of an undamped current in thermodynamical equilibrium — the so-called persistent current — flowing in an isolated normal mesoscopic ring, pierced by a magnetic flux [Moskalets 2010]. This persistent current, similar to the conductance above, is the periodic function of the magnetic flux. This current is obtained by summing up the partial currents of different states with the quantum number n, which comes from the periodic boundary conditions for the wave function.

   Let us calculate the persistent current in a one-dimensional isolated ring. Assume either that there is a solenoid inside this ring, or the ring is placed in the weak magnetic field, so that the ring is pierced by the magnetic flux Φ, and the Zeeman term in the Hamiltonian can be neglected. Consider the coordinate x along the ring. Since the circulation of the vector potential gives the magnetic flux, we have $A_x = -A = -\Phi/L$. Then the Hamiltonian takes the form:

$$H = \frac{1}{2m}\left(-i\hbar\frac{\partial}{\partial x} - \frac{e}{c}A_x\right)^2 = -\frac{\hbar^2}{2m}\left(\frac{\partial}{\partial x} - i\frac{2\pi\Phi}{\Phi_0}\frac{1}{L}\right)^2,$$

$$\Phi_0 = \frac{hc}{|e|}. \tag{4.25}$$

For the solution of the Schrödinger equation $H\psi = E\psi$ we obtain

$$\psi = \frac{1}{\sqrt{L}}\exp\left(ikx + i\frac{2\pi\Phi}{\Phi_0}\frac{x}{L}\right), \qquad \int_0^L dx\,|\psi(x)|^2 = 1,$$

$$E = \frac{\hbar^2 k^2}{2m}. \tag{4.26}$$

The wave vector k is defined from the single-valuedness condition for the wave function, $\psi(x + L) = \psi(x)$, from which we have $kL + 2\pi\Phi/\Phi_0 = 2\pi n$, or

$$k_n = \frac{p_n}{\hbar} = \frac{2\pi}{L}\left(n - \frac{\Phi}{\Phi_0}\right). \tag{4.27}$$

From here, the expression for the spectrum follows,

$$E_n = \frac{h^2}{2mL^2}\left(n - \frac{\Phi}{\Phi_0}\right)^2. \tag{4.28}$$

Let us calculate the partial current of an electron:

$$I_n = e\left(\psi_n^* \frac{\hat{p} - eA/c}{2m}\psi_n + c.c.\right)$$

$$= \frac{e\hbar}{2m}2\mathrm{Re}\left(\psi_n^*\left(-i\frac{\partial}{\partial x} - \frac{2\pi\Phi}{\Phi_0}\frac{1}{L}\right)\psi_n\right) = \frac{e\hbar k_n}{mL}, \tag{4.29}$$

or

$$I_n = \frac{eh}{mL^2}\left(n - \frac{\Phi}{\Phi_0}\right). \tag{4.30}$$

Note, first, that $I_{-n} \neq I_n$ and, second, that $I_n = -\frac{1}{c}\frac{\partial E_n}{\partial \Phi}$.

In order to obtain the total current, we need to sum up the partial currents while accounting for the distribution function:

$$I = \sum_{n=-\infty}^{\infty} I_n f_F(E_n) = \frac{eh}{mL^2} \sum_{n=-\infty}^{\infty} \left(n - \frac{\Phi}{\Phi_0}\right) f_F(E_n). \tag{4.31}$$

Consider for simplicity the case of zero temperature. Then the Fermi distribution function indicates that all the states under the Fermi level, $E_n < \mu$, are occupied, and above the Fermi level there are no occupied states. Then there are two possible cases — when the number of occupied states is odd or even, see Fig. 4.6.

Consider, first, the case, where the total number of electrons N is *odd*. Then in Eq. (4.31) we have

$$\sum_{n=-(N-1)/2}^{(N-1)/2} n = 0, \qquad \sum_{n=-(N-1)/2}^{(N-1)/2} 1 = N. \tag{4.32}$$

And also let us go from the total electron number to the Fermi velocity by noting that in Eq. (4.27) we have $p_F = \frac{2\pi\hbar}{L}N$, then it

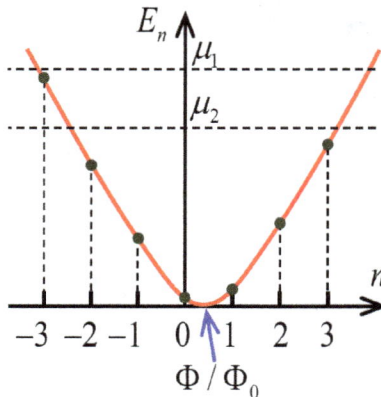

Fig. 4.6. Occupied energy states for two possible values of the chemical potential.

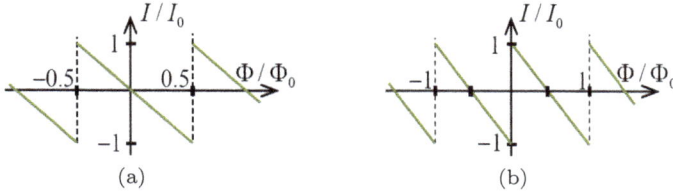

Fig. 4.7. Dependence of the persistent current in the ring on the magnetic flux for the cases of odd (a) and even (b) number of electrons in the ring.

follows that $v_F = \frac{hN}{mL}$. And we obtain

$$I_{\text{odd}} = -I_0 \frac{\Phi}{\Phi_0}, \qquad \left| \frac{\Phi}{\Phi_0} \right| < \frac{1}{2}. \tag{4.33}$$

Here, the current amplitude $I_0 = ev_F/L$ corresponds to the current of one electron at the Fermi level in the ring. The obtained saw-tooth dependence of the current on the magnetic flux is shown in Fig. 4.7(a). This is the diamagnetic current.

For the case of *even* number of electrons in the ring: $\sum n = \frac{N}{2}\text{sgn}\Phi$ and for the positive value of the flux we obtain

$$I_{\text{even}} = I_0 \left(1 - 2\frac{\Phi}{\Phi_0} \right), \qquad 0 < \frac{\Phi}{\Phi_0} \leq 1. \tag{4.34}$$

The obtained dependence is demonstrated in Fig. 4.7(b). We note that, for even N, the dependence becomes paramagnetic.

We emphasize that the change in the number of electrons by 1 changes the ring response — from diamagnetic to paramagnetic. This kind of effect is called the *parity effect*.

* Note that the obtained formulas, Eqs. (4.33) and (4.34), can be unified in the form of the Fourier expansion

$$I = I_0 \frac{2}{\pi} \sum_{k=1}^{\infty} (-1)^{kN} \frac{\sin(2\pi k\Phi/\Phi_0)}{k}. \tag{4.35}$$

* At arbitrary non-zero temperature, the summation in Eq. (4.31) can be transformed to the form of a Fourier series by making use of

the known and useful Poisson summation formula:

$$\sum_{n=-\infty}^{\infty} g(n) = \int_{-\infty}^{\infty} dx g(x) + 2\mathrm{Re} \sum_{k=1}^{\infty} \int_{-\infty}^{\infty} dx g(x) \, e^{i2\pi kx}. \qquad (4.36)$$

As the result, one can obtain [Moskalets 2010]:

$$I = I_0 \frac{2}{\pi} \frac{T}{T^*} \sum_{k=1}^{\infty} \frac{\cos\left(2\pi k L/\lambda_F\right)}{\sinh\left(kT/T^*\right)} \sin\left(2\pi k \Phi/\Phi_0\right), \quad T^* = \frac{h v_F}{2\pi^2 L}.$$

$$(4.37)$$

(We note that $2\pi/\lambda_F = k_F = 2\pi N/L$, if the Fermi level either coincides with one of the doubly-degenerate levels or hits exactly in the middle between them.) From here, in particular, at low temperatures, we obtain the formula (4.35). At high temperatures, the series can be limited by the first term and then it can be seen that the effect exponentially decreases as the temperature increases.

So, we can see that the persistent current has a number of interesting properties: (i) it is periodic in the flux Φ, (ii) its amplitude I_0 is defined by the contribution of a single electron at the Fermi level, (iii) it displays the parity effect, (iv) it decays exponentially at $T > T^*$.

4.4. Coulomb blockade

Above we have considered the peculiarities of quantum transport through a quantum wire and a doubly-connected conductor pierced by the magnetic flux. Consider now the quantum transport through a quantum dot, of which the schematic is shown in Fig. 4.8. A *quantum dot* (also an island or a grain) is formed by several capacitances C_j, to which the electric potentials φ_j are applied. The gate voltage V_g is used to control the quantum-dot state and, correspondingly, the transport current, created by the bias voltage V. Note that the superconducting analogue of a quantum dot — the Cooper-pair box — was considered in Sec. 3.4.

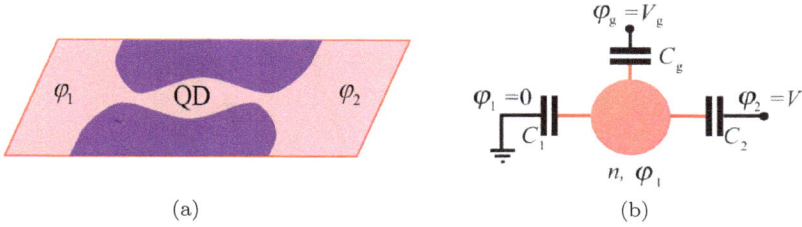

Fig. 4.8. (a) Schematic of the quantum dot (QD) weakly coupled to the 2DEG and biased by the electric potentials $\varphi_{1,2}$. (b) Equivalent schematic of the quantum dot, to which the bias voltage $V = \varphi_2 - \varphi_1$ and the gate voltage V_g are applied.

4.4.1. *Energy levels in a quantum dot*

Let us calculate the energy associated with adding n electrons to an initially electro-neutral island. The added charge equals the sum of the charges on the inner plates of the capacitors, to which the potential difference $\Delta\varphi_j = \varphi_I - \varphi_j = \varphi_I - V_j$ is applied (φ_I stands for the island potential):

$$ne = \sum C_j \Delta\varphi_j = \varphi_I \sum C_j - \sum C_j V_j \equiv C_\Sigma \varphi_I + e n_g,$$

$$(4.38)$$

$$C_\Sigma = \sum C_j, \quad n_g = -\frac{1}{e} \sum C_j V_j.$$

From here, the expression for the island potential follows: $\varphi_I = e(n - n_g)/C_\Sigma$. Then for the electrostatic energy of the island we get

$$E = \sum \frac{C_j \varphi_I^2}{2} = \frac{e^2(n - n_g)^2}{2C_\Sigma^2} \sum C_j = E_C(n - n_g)^2, \qquad (4.39)$$

where $E_C = e^2/2C_\Sigma$ is the characteristic charging energy of the island. If we minimize the energy (4.39), then we obtain $n_0 = n_g$, and this is the so-called induced or gate charge. Since the number of electrons is an integer, this corresponds to the number in the interval $n_0 - \frac{1}{2} \leq n \leq n_0 + \frac{1}{2}$. In order to demonstrate this, the energy levels $E/E_C = (n - n_g)^2$ are plotted in Fig. 4.9 as a function of the gate charge n_g, that is, as a matter of fact, the dimensionless voltage.

Let us analyze the system energy levels at $V = 0$, where we are interested in the dependence on $n_g = C_g V_g/|e|$. We can see that

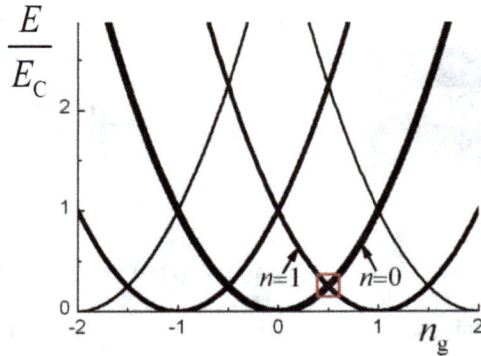

Fig. 4.9.   Energy levels, E/E_C, in a quantum dot for several values of the excess number of electrons n.

at the integer values of the parameter n_g, the energy levels, which correspond to the change in the number of electrons on the island, are maximally separated from one another. And in this case, in order to change the island charge by the charge of one electron, we need to change the system energy by $E = E_C$. At integer n_g, the application of a small bias voltage cannot change the island charge, and the current through it would not pass. This regime is called the *Coulomb blockade*. At $|V| < E_C/|e|$, the current through the island equals zero; at $|V| > E_C/|e|$ the current, linear in voltage, appears, which is related to the resonant tunneling of single electrons — at integer n_g.

If n_g is a half-integer, say $n_g = 1/2$, then changing the island charge does not require energy: the states with $n = 0$ and $n = 1$ have equal energies — and the island is *deblockaded*. At vanishingly small bias voltage, the current will flow through the quantum wire. Such a device is called a *single-electron transistor* (SET). The field involving the study of such devices, for which the Coulomb blockade is important, is sometimes called *single-electronics*.

4.4.2. *Conductance of a quantum dot*

The state of a quantum dot is characterized by the probability P_n of n electrons dwelling in it. The set of these probabilities can be described by the balance equation (which is also called the kinetic or master equation). For this, let us define the value $\Gamma_{n+1,n}$ to be the number of transitions per unit time (transition rate) from the state n to the state $n+1$. Assuming that only transitions between neighboring states are possible, we can write:

$$\frac{dP_n}{dt} = \Gamma_{n,n+1}P_{n+1} + \Gamma_{n,n-1}P_{n-1} - \left(\Gamma_{n+1,n} + \Gamma_{n-1,n}\right)P_n.$$
$$(4.40)$$

The kinetic equation (4.40) for the stationary distribution is equivalent to the detailed balance equation. Indeed, this equation, with the l.h.s. zero, is fulfilled under the condition

$$\Gamma_{n,n+1}P_{n+1} = \Gamma_{n+1,n}P_n. \qquad (4.41)$$

Now, let us calculate the current through the quantum dot. Let $\Gamma_L^\pm(n)$ and $\Gamma_R^\pm(n)$ denote the transition rates from the left (L) and right (R) electrodes to (+) and from (−) the dot, see Fig. 4.10.

Then we have

$$\Gamma_{n+1,n} = \Gamma_L^+(n) + \Gamma_R^+(n),$$
$$\Gamma_{n-1,n} = \Gamma_L^-(n) + \Gamma_R^-(n), \qquad (4.42)$$

and for the current, say, between the left electrode and the dot:

$$I = e\sum_{n=0}^{\infty} P_n \left(\Gamma_L^+(n) - \Gamma_L^-(n)\right). \qquad (4.43)$$

Consider the conductivity for a dot, in which only the two states, with N and $N+1$ electrons, are important. Denote these states as 0

Fig. 4.10.   Tunneling of electrons into the quantum dot and out of it.

and 1, which refer to the number of excess electrons. Then we need to consider only two probabilities P_0 and P_1, for which $P_0 + P_1 = 1$. Then from the detailed-balance equation (4.41) we have

$$P_1 = \frac{\Gamma_{1,0}}{\Gamma_\Sigma}, \quad P_0 = \frac{\Gamma_{0,1}}{\Gamma_\Sigma}, \quad \Gamma_\Sigma = \Gamma_{1,0} + \Gamma_{0,1}. \qquad (4.44)$$

The rates of transitions between the two states in the dot have the form:

$$\Gamma_{1,0} = \Gamma_L^+(0) + \Gamma_R^+(0), \quad \Gamma_{0,1} = \Gamma_L^-(1) + \Gamma_R^-(1). \qquad (4.45)$$

Putting (4.45) in Eq. (4.44), we obtain the current from Eq. (4.43)

$$I = e\frac{\Gamma_R^-(1)\Gamma_L^+(0) - \Gamma_R^+(0)\Gamma_L^-(1)}{\Gamma_\Sigma}. \qquad (4.46)$$

In order to proceed further, we note that the transition rate is related to the occupation of the respective state in the banks, that is the Fermi distribution function:

$$\Gamma_{L,R}^+(0) = \Gamma_0^{L,R} f_F(\varepsilon_0 - \mu_{L,R}), \quad \mu_{L,R} = \mu \pm \frac{eV}{2}, \qquad (4.47)$$

$$\Gamma_{L,R}^-(0) = \Gamma_0^{L,R} \left[1 - f_F(\varepsilon_0 - \mu_{L,R})\right]. \qquad (4.48)$$

Here ε_0 is the energy of the state with 0 excess electrons in the dot. Also, for convenience, we assumed that the voltage is applied symmetrically, as $\pm eV/2$ from the left and from the right. Analogously, we can write down the rates for the state with 1 excess electron. For simplicity, consider the case of the deblockaded dot, when $\varepsilon_1 = \varepsilon_0$. Then $\Gamma_{L,R}^\pm(1) = \Gamma_{L,R}^\pm(0) \equiv \Gamma_{L,R}^\pm$. Also we remind the reader that in the approximation linear in eV/μ, we have for the Fermi function

$$f_F\left(\varepsilon - \mu \mp \frac{eV}{2}\right) \approx f_F(\varepsilon - \mu) \mp \frac{eV}{2} f_F'(\varepsilon - \mu), \qquad (4.49)$$

$$f_F' = -\frac{1}{4k_BT}\cosh^{-2}\frac{\varepsilon - \mu}{2k_BT} \xrightarrow[T\to 0]{} -\delta(\varepsilon - \mu). \qquad (4.50)$$

Then in the approximation linear in voltage:

$$\Gamma_R^- \Gamma_L^+ - \Gamma_R^+ \Gamma_L^- = \Gamma_0^R \Gamma_0^L \left(f_F(\varepsilon - \mu_L) - f_F(\varepsilon - \mu_R) \right)$$

$$\approx \Gamma_0^R \Gamma_0^L eV(-f_F'), \quad \Gamma_\Sigma \approx \Gamma_0^L + \Gamma_0^R. \quad (4.51)$$

Let us also introduce $\tilde{\Gamma}_{L,R} = h\Gamma_0^{L,R}$, and then for the conductance $G = I/V$ we obtain

$$G = G_0 \frac{\tilde{\Gamma}_L \tilde{\Gamma}_R}{\tilde{\Gamma}_L + \tilde{\Gamma}_R} \frac{1}{4k_B T} \cosh^{-2} \frac{\varepsilon - \mu}{2k_B T}, \quad G_0 = \frac{2e^2}{h}. \quad (4.52)$$

This formula has to be related to the classical case of two parallel resistances, which looks as follows:

$$R = \frac{1}{G} = R_L + R_R = \frac{1}{G_L} + \frac{1}{G_R} = \frac{G_L + G_R}{G_L G_R} \;\; \Rightarrow \;\; G = \frac{G_L G_R}{G_\Sigma}.$$
$$(4.53)$$

In particular, for $G_L = G_R$, it follows that $G = G_L/2$.

 * A separate interest also presents the derivation of the formula (4.47): $\Gamma_j^+(0) = \Gamma_0^j f_F(\varepsilon - \mu_j)$. In order to obtain the transition frequencies, one needs to use Fermi's golden rule, see Eq. (2.117):

$$\Gamma_{i \to f} = \frac{2\pi}{\hbar} |\langle f | H_t | i \rangle|^2 \delta(\varepsilon_f - \varepsilon_i), \quad (4.54)$$

where the tunneling Hamiltonian

$$H_t = t_j \sum_k a_k^{(j)} c^\dagger + h.c., \quad (4.55)$$

is described by the amplitude t_j, the annihilation operator of an electron in the j-th electrode $a_k^{(j)}$, and the creation operator in the dot $c^\dagger$.

Conclusion to Chapter 4

Mesoscopic-size conductors have a number of basic differences from macroscopic conductors. For them the charge and spectrum quantization are important.

The conductance of a mesoscopic conductor has a step-wise dependence on its width, which is distinct from the linear dependence for a macroscopic conductor. Besides, the conductance is defined by the probability of transmission through the conductor. If the transmission is ballistic, then the height of the steps is defined by the conductance quantum, $G_0 = 2e^2/h$. If the conductor has one conducting channel (i.e. its width is of the order of the Fermi wavelength), then, even in the absence of impurities, its resistance has finite value, $R_0 = 1/G_0 \approx 13 \ k\Omega$.

For mesoscopic conductors, it is important to account for possible interference. In particular, for a doubly-connected conductor, the Aharonov–Bohm effect is displayed as the oscillatory dependence of the conductance on the magnetic flux, with the periodicity $\Phi_0^N = h/|e| \approx 4 \cdot 10^{-15}$ Wb.

The conductance of a mesoscopic-size island is described by the Coulomb-blockade regime. The electrons tunnel one by one and the gate voltage can control the current through such a single-electron transistor.

Problems for independent work and for self-assessment

4.1. (**) Find the eigen-values and the eigen-functions of a free particle in a 3D box.

4.2. (***) Derive the Landauer formula for the conductance of a 1D conductor with a single impurity; note the difference from Ohm's law.

4.3. (***) Plot the conductance of a quantum wire, described by the Landauer formula, for non-zero temperature to demonstrate how it washes out the steps.

4.4. (**) Demonstrate that the Aharonov–Bohm effect results in the periodic dependence of conductance on the magnetic flux, with the period equal to the magnetic flux quantum.

4.5. (*****) Calculate the transmission coefficient for the doubly-connected ballistic conductor; see Eq. (4.24) and the reference next to it.

4.6. (**) Write down the Hamiltonian for an electron in a 1D ballistic ring; find its eigen-energies and eigen-states.

4.7. (***) After the previous task, write down and plot the dependence of the current on the magnetic flux in the ring for the two cases, where their number is odd and even, respectively.

4.8. (*****) Calculate the persistent current in a ring for non-zero temperature; see Eq. (4.37) and the reference next to it.

4.9. (*) Analyze the persistent current, Eq. (4.37), at high temperatures.

4.10. (**) Analyze the spectrum of electrons on a quantum dot and describe the Coulomb blockade and a single-electron transistor.

4.11. (****) Study the conductance of a quantum dot in terms of the transition rates.

Chapter 5

CIRCUIT QUANTUM ELECTRODYNAMICS

"The long journey has passed,
After the distant cloud.
I will sit to rest."

Matsuo Basho[34]

As discussed previously, it is interesting and important to study not only single mesoscopic systems, but also their connection with other subsystems (see Footnote 9 on page 15). Below, we will consider two illustrative examples, one on the system of a quantum dot coupled to a classical nanomechanical oscillator and another for the system composed of a superconducting qubit and a quantum resonator on the base of a transmission line. In the former problem, we will consider the case of a slow resonator, of which the frequency ω_0 is much smaller than all other characteristic frequencies in the problem. In particular, if the respective distance between the quantum levels is much smaller than the thermal broadening, $\hbar\omega_0 \ll k_B T$, then such a resonator should be treated as a classical one. And the system "a quantum dot — a classical resonator" will be considered in the framework of the so-called semi-classical theory. On the latter example we will consider the inverse situation, when

---

[34]Matsuo Basho. This haiku was translated by the author from "Japanese lyric poetry. Matsuo Basho," AST publishers, Moscow, translated by V. Sokolov (2002). {This haiku in Russian: "Долгий путь пройден, За далеким облаком. Сяду отдохнуть."}

$\hbar\omega_0 > k_B T$. In this case, the resonator should be considered as a quantum-mechanical object and the whole system "qubit-resonator" is described in full analogy with an atom interacting with a quantum field. Making such an analogy, we will familiarize ourselves with the formalism to describe such systems.

So, the system "qubit-resonator" is analogous to one of the basic systems of quantum optics — an atom in an electromagnetic field. And here it is important to note that the large number of effects for such a system is described by the semi-classical theory, where an atom is treated as a discrete quantum system, and the field is assumed classical.[35] In the general case, quantum consideration is needed, which implies using the elements of the quantum optics theory [Scully and Zubairy 2012].

5.1. Quantum dot and nanomechanical oscillator

Consider a quantum dot (island) in such a configuration as in the last section of the previous chapter with only one distinction — one of the capacitor plates is now a suspended bridge, which can oscillate (see Fig. 5.1). Such an object of small cross-section is called a *nanomechanical resonator*. If its frequency is small in comparison with the other characteristic frequencies, then it is described as a classical oscillator. Of course, especially interesting is the situation when such an object can pass to the quantum regime. Here we will not consider this situation but rather we will consider this in the next section, on the example of another resonator.

So, here we consider a normal island, created by the three capacitances C_1, C_2, and C_{NR}. One of the plates of the capacitor C_{NR} is formed by the nanomechanical resonator, or, for brevity, the nanoresonator. The position of the nanoresonator is characterized by the displacement of its central point u. This displacement is assumably much smaller than the distance d between the plates. Then the capacitance between the nanoresonator and the qubit has

---

[35] N. B. Delone and V. P. Krainov, Atoms in Strong Light Fields, Springer Series in Chemical Physics Vol. 28, Springer, Berlin (1985); Atomizdat, Moscow (1978).

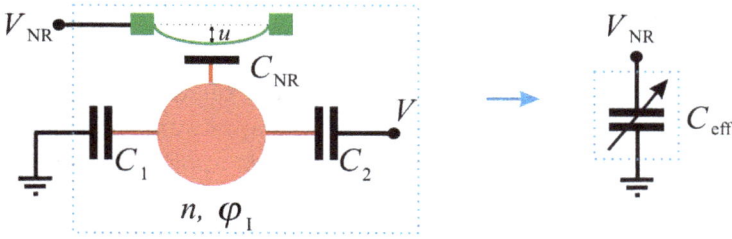

Fig. 5.1. Quantum dot and nanomechanical resonator; description of the system by introducing effective (quantum) capacitance C_{eff}.

the form

$$C_{\text{NR}}(u) \approx C_{\text{NR0}} + \left.\frac{\partial C_{\text{NR}}}{\partial u}\right|_{u=0} u \equiv C_{\text{NR}}\left(1 + \frac{u}{\xi}\right),$$

$$\xi^{-1} = \frac{1}{C_{\text{NR0}}}\left.\frac{\partial C_{\text{NR}}}{\partial u}\right|_{u=0}, \quad \xi \sim d \gg u. \tag{5.1}$$

By the subscript "0" we have here denoted the value at $u = 0$; hereafter we will not write down this index. In order to make the estimates, we can consider the capacitor to be plane-parallel for which

$$C_{\text{NR}}(u) = \frac{\varepsilon\varepsilon_0 S}{d + u} \approx C_{\text{NR0}} - \frac{C_{\text{NR0}}}{d}u, \tag{5.2}$$

from where, in particular, we can see that $\xi = -d$. Here ε stands for the dielectric constant of the medium, equal to 1 for the vacuum; $\varepsilon_0 \approx 8.8 \cdot 10^{-12}$ F/m is the electric constant.

The displacement of the nanoresonator influences the qubit via the changes of the polarization charge. (In order to have this influence essential, they apply large voltage V_{NR}, of the order of several volts.) Also, we assume that there is the electronics, which can probe the oscillation frequency of the nanoresonator, which, for simplicity, we do not show in Fig. 5.1. Let us aim to define the state (i.e. the charge) of the quantum island by means of measuring the frequency of the classical nanoresonator.

Consider a *semi-classical theory* for our system, when the island is described as a quantum system and the resonator is described as a classical system. Let the force, acting on the nanomechanical resonator, drive it from the side of the quantum subsystem, the quantum dot, and also assume that there is another periodic probe force $F_p \sin \omega_p t$. Then we have the equation for the displacement u of the classical nanoresonator with the effective mass m and the eigen-frequency ω_0, under the influence of the force, and without accounting for the dissipation:

$$m\frac{d^2u}{dt^2} + m\omega_0^2 u = F_q\left(u, \frac{du}{dt}\right) + F_p \sin \omega_p t. \qquad (5.3)$$

Note here that oscillations in a non-linear system, described by Eq. (5.3), can be reduced to the oscillations of an equivalent linear system making use of the *Krylov–Bogolyubov asymptotic expansion* formalism.[36] Here we would like to demonstrate how to act even simpler: we will consider the system, composed of a quantum dot and a classical resonator by introducing the *parametric* (or, as they say, *quantum*) *capacitance*.

The change of the number of electrons on the island is related to their stochastic tunneling. But, since the resonator frequency is small, it "sees" the average island occupation $\langle n \rangle$. As we have discussed in the previous section (see Eq. (4.38)), the island potential is related to the average number of electrons on the island,

$$\varphi_I = \frac{e\left(\langle n \rangle - n_g\right)}{C_\Sigma},$$

$$C_\Sigma = \sum C_j, \quad n_g = -\frac{1}{e}\sum C_j V_j = -\frac{C_2 V + C_{NR} V_{NR}}{e}. \qquad (5.4)$$

Then, for the charge we have $Q_{NR} = (V_{NR} - \varphi_I)C_{NR}$ and we can define the effective differential capacitance, as it is shown in Fig. 5.1,

[36] N. N. Bogoliubov and Yu. A. Mitropolsky, Asymptotic Methods in the Theory of Non-linear Oscillations (Gordon and Breach, New York, 1961); Nauka, Moscow (1974).

by differentiating the charge Q_{NR} on the capacitor (C_{NR}) plate:

$$C_{\mathrm{eff}}(u) = \frac{\partial Q_{\mathrm{NR}}}{\partial V_{\mathrm{NR}}} = C_{\mathrm{NR}} - \frac{eC_{\mathrm{NR}}}{C_{\Sigma}} \left(\frac{\partial \langle n \rangle}{\partial V_{\mathrm{NR}}} + \frac{\partial}{\partial V_{\mathrm{NR}}} \frac{C_{\mathrm{NR}} V_{\mathrm{NR}}}{e} \right)$$

$$\equiv C_{\mathrm{geom}} + C_{\mathrm{Q}},$$

$$C_{\mathrm{geom}} = C_{\mathrm{NR}} - \frac{C_{\mathrm{NR}}^2}{C_{\Sigma}} = \frac{C_{\mathrm{NR}}(C_1 + C_2)}{C_{\Sigma}} \approx C_{\mathrm{NR}}, \quad C_{\mathrm{NR}} \ll C_{1,2},$$

$$C_{\mathrm{Q}}(u) = -\frac{eC_{\mathrm{NR}}}{C_{\Sigma}} \frac{\partial \langle n \rangle}{\partial V_{\mathrm{NR}}} = \frac{C_{\mathrm{NR}}^2}{C_{\Sigma}} \frac{\partial \langle n \rangle}{\partial n_{\mathrm{g}}}. \tag{5.5}$$

Here, the effective capacitance was split into geometric and (small) "quantum" constituents.

Now we can consider the force, which acts on the nanoresonator from the side of the island, as an electrostatic force with the effective capacitance C_{eff}. Beside the direct voltage, let there be a small alternating voltage applied to the nanoresonator:

$$V_{\mathrm{NR}}(t) = V_{\mathrm{NR}} + V_A \sin \omega_{\mathrm{p}} t, \quad V_A \ll V_{\mathrm{NR}}. \tag{5.6}$$

Then, we take into consideration that $V_{\mathrm{NR}}(t)^2 \approx V_{\mathrm{NR}}^2 + 2 V_{\mathrm{NR}} V_A \sin \omega_{\mathrm{p}} t$, $C_{\mathrm{NR}}(u)^2 \approx C_{\mathrm{NR0}}^2 (1 + u/\xi)^2$, and that accounting for the dependence on u in the other terms result in negligibly small corrections, and we obtain

$$F_{\mathrm{q}} = \frac{1}{2} \frac{\partial}{\partial u} \left(C_{\mathrm{eff}}(u) V_{\mathrm{NR}}^2(t) \right) \approx \frac{1}{2} V_{\mathrm{NR}}^2 \frac{\partial (C_{\mathrm{NR}} + C_{\mathrm{Q}})}{\partial u}$$

$$+ V_{\mathrm{NR}} V_A \sin \omega_{\mathrm{p}} t \frac{\partial C_{\mathrm{NR}}}{\partial u} \tag{5.7}$$

$$= F_0 + F_{\mathrm{q}}(u) + F_{\mathrm{p}} \sin \omega_{\mathrm{p}} t.$$

Here we have defined the amplitude $F_{\mathrm{p}} = V_{\mathrm{NR}} V_A \frac{C_{\mathrm{NR}}}{\xi}$, and separated the displacement-dependent term

$$F_{\mathrm{q}} = V_{\mathrm{NR}}^2 C_{\mathrm{Q}} \frac{u}{\xi^2}, \quad C_{\mathrm{Q}} = \frac{C_{\mathrm{NR}}^2}{C_{\Sigma}} \frac{\partial \langle n \rangle}{\partial n_{\mathrm{g}}}, \tag{5.8}$$

and we denoted the (insignificant) constant terms with F_0. If we move the term F_q in the equation of motion (5.3) to the left side of the equality, we obtain the shift for the term linear in the displacement u:

$$m\omega_0^2 \rightarrow m\omega_0^2 - \frac{V_{NR}^2}{\xi}C_Q = m\omega_{\text{eff}}^2 = m(\omega_0 + \Delta\omega)^2$$

$$\approx m\omega_0^2 + 2m\omega_0\Delta\omega. \tag{5.9}$$

This means that for the frequency shift we obtain

$$\Delta\omega = -\frac{V_{NR}^2}{2m\omega_0\xi}C_Q \propto \frac{\partial\langle n\rangle}{\partial n_g}. \tag{5.10}$$

So, the frequency shift is defined by the quantum capacitance. Experimentally, this allows the state of the quantum subsystem (here, the quantum dot) to be defined by measuring the resonator frequency shift (here, the nanomechanical resonator).

* *Quantum dot–charge qubit.* In conclusion, consider the situation where only two states are relevant for the quantum dot, which means that it operates as a charge qubit. Then, it is not difficult to show that the ground and excited states correspond to the quantum capacitances of opposite signs. Indeed, as we have discussed, such a charge qubit is described by the Hamiltonian

$$H = -\frac{\varepsilon_0}{2}\sigma_z - \frac{\Delta}{2}\sigma_x, \quad \varepsilon_0 = 2E_C\left(n_g - n_g^{(0)}\right), \tag{5.11}$$

where the value Δ defines the tunneling between the two charge states. By linking the charge basis and eigen-energy basis, we have

$$\langle n\rangle = P_-\langle n\rangle_- + P_+\langle n\rangle_+ = \langle n\rangle_- + P_+\left(\langle n\rangle_+ - \langle n\rangle_-\right),$$

$$\langle n\rangle_\pm = \frac{1}{2}\left(1 \pm \frac{\varepsilon_0}{\Delta E}\right), \quad \Delta E = \sqrt{\Delta^2 + \varepsilon_0^2}. \tag{5.12}$$

In the ground/excited states we have $P_\pm = 0$, from where $\langle n\rangle = \langle n\rangle_\pm$ and

$$\frac{\partial\langle n\rangle}{\partial n_g} = \pm\frac{\partial}{\partial n_g}\frac{\varepsilon_0}{2\Delta E}. \tag{5.13}$$

From here it can be seen that the quantum capacitances and the frequency shifts (5.10) in the ground and excited states have opposite signs.

5.2. Flux qubit and transmission-line resonator

In this section we will consider the system "qubit-resonator" in the case where the *resonator is in the quantum regime*. For concreteness, we consider the system composed of a flux superconducting qubit inductively coupled to the transmission-line resonator. We will demonstrate the remarkable fact that such a system is analogous to an atom in electromagnetic field.

First, we consider the realization of the quantum resonator on the base of the transmission line. Such a resonator is situated between the two cuts in the transmission line, which form the capacitors C_0 and are situated at $x = \pm l/2$, see Fig. 5.2. We assume that the qubit is inductively coupled to the current in the resonator and is situated at its center, $x = 0$.

Let us start from describing a transmission line.

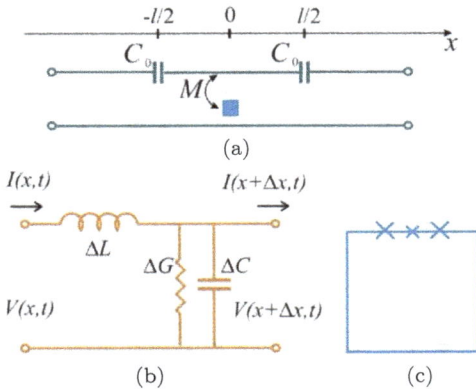

Fig. 5.2. (a) Schematic of the qubit inductively coupled to the quantum resonator on the base of a transmission line; (b) equivalent circuit for the description of a small section of the transmission line; (c) the flux qubit with the three Josephson junctions.[37]

---

[37] A. N. Omelyanchouk, E. V. Il'ichev, and S. N. Shevchenko, Quantum coherent phenomena in Josephson qubits (Naukova Dumka, Kiev, 2013).

5.2.1. *Transmission line*

A transmission line can be described as an infinite series of the elementary circuits, as it is shown in Fig. 5.2(b).[38] Here the elementary inductance, capacitance, and conductance are equal to: $\Delta L = L\Delta x$, $\Delta C = C\Delta x$, and $\Delta G = G\Delta x$, where L, C and G stand for the inductance, capacitance, and conductance (which describes the dielectric losses in the material between the two conductors) per unit length. Neglecting the Ohmic losses ($R = 0$), we can write down the following equations, using the Kirchhoff laws for the current $I(x,t)$ and the voltage $V(x,t)$:

$$I(x,t) - I(x + \Delta x, t) = \Delta G V(x,t) + \Delta C \frac{\partial V(x,t)}{\partial t},$$

$$V(x,t) - \Delta L \frac{\partial I(x,t)}{\partial t} - V(x + \Delta x, t) = 0. \tag{5.14}$$

Dividing by Δx, and tending Δx to zero, we obtain the well-known *telegraph equations*:

$$\frac{\partial I(x,t)}{\partial x} = -GV(x,t) - C\frac{\partial V(x,t)}{\partial t},$$

$$\frac{\partial V(x,t)}{\partial x} = -L\frac{\partial I(x,t)}{\partial t}. \tag{5.15}$$

These equations can be written identically for the current and the voltage:

$$\left(\frac{\partial^2}{\partial x^2} - \frac{1}{v^2}\frac{\partial^2}{\partial t^2} + \frac{\kappa}{v^2}\frac{\partial}{\partial t} \right) \begin{Bmatrix} I(x,t) \\ V(x,t) \end{Bmatrix} = 0, \tag{5.16}$$

where $v = 1/\sqrt{LC}$ has the meaning of the phase velocity and $\kappa = G/C$ describes the losses in the transmission line.

Consider further the monochromatic wave, for which

$$I(x,t) = I(x)e^{i\omega t}, \quad V(x,t) = V(x)e^{i\omega t}, \tag{5.17}$$

---

[38]D. M. Pozar, Microwave Engineering, Wiley, New York (1990).

where taking the real part is implied. Such substitution gives

$$\left(\frac{\partial^2}{\partial x^2} + \gamma^2\right) V(x) = 0, \tag{5.18}$$

where we have defined the value γ, which can be written as follows, accounting for the smallness of the losses:

$$\gamma = \sqrt{\frac{\omega^2}{v^2} + i\frac{\kappa\omega}{v^2}} = \frac{\omega}{v}\sqrt{1 + i\frac{\kappa}{\omega}} \approx \frac{\omega}{v} + i\frac{\kappa}{2v} \equiv k + i\alpha. \tag{5.19}$$

We obtain the solution of Eq. (5.18) for the voltage

$$V(x) = V_+ e^{i\gamma x} + V_- e^{-i\gamma x}. \tag{5.20}$$

Note that e^{-ikx} corresponds to the wave propagating to the right and e^{ikx} corresponds to the wave propagating to the left.

From Eq. (5.15) we have the relation between the current and the voltage

$$\frac{\partial V(x)}{\partial x} = -i\omega L I(x) \Rightarrow I(x) = \frac{i}{\omega L} i\gamma (V_+ e^{i\gamma x} - V_- e^{-i\gamma x})$$

$$\approx -\frac{k}{\omega L}(V_+ e^{i\gamma x} - V_- e^{-i\gamma x}), \tag{5.21}$$

or

$$I(x) = -\frac{V_+}{Z_0} e^{i\gamma x} + \frac{V_-}{Z_0} e^{-i\gamma x}, \quad Z_0 = \sqrt{\frac{L}{C}}. \tag{5.22}$$

Here the value Z_0 denotes the transmission-line impedance.

5.2.2. *Transmission-line resonator*

Consider an open transmission line of length l, forming the resonator. Let us now define normal modes of the resonator without dissipation ($\kappa = 0$). Then, assuming that the current through the boundaries at $x = \pm l/2$ equals to zero, we obtain

$$V_+ = -V_-, \quad k_j \frac{l}{2} = \pi j - \frac{\pi}{2}, \quad j = 1, 2, \dots,$$

$$I_j(x) = \frac{2V_-}{Z_0} \cos k_j x, \quad V_j(x) = -i2V_- \sin k_j x. \tag{5.23}$$

In particular, for the fundamental mode of the $\lambda/2$-resonator ($l = \lambda/2$) we have $k_r \equiv k_1 = \pi/l$ and $\omega_r \equiv \omega_1 = k_1 v = \pi/\sqrt{L_r C_r}$, where $L_r = Ll$ and $C_r = Cl$ are the full inductance and capacitance of the resonator.

Further we can expand the current in the resonator over the normal modes. For simplification of the situation, we consider the frequencies close to the frequency ω_r, so that we can ignore all the modes besides the fundamental one. We have specified the coordinate dependence and now we return to the time-dependent expressions for the current and voltage for the fundamental mode with $k_1 = \pi/l$ (in what follows we omit the subscript $j = 1$)

$$I(x,t) = Aq(t) \cos kx, \tag{5.24}$$

where we describe the time dependence by the product $Aq(t)$, in which the value q will be chosen later as a generalized coordinate and the constant A will be chosen from the considerations of making the analogy with a harmonic oscillator. Then for the voltage we obtain

$$V(x,t) = -L \int_0^x dx' \frac{\partial I(x',t)}{\partial t} = -\frac{LA}{k} \dot{q}(t) \sin kx. \tag{5.25}$$

Next, we introduce the Hamiltonian as the total energy of the resonator, which is better written by choosing $A = \sqrt{\frac{2m}{L_r}} \omega_r$, with m standing for some multiplier of the dimensionality of mass:

$$H_r = \int_{-l/2}^{l/2} dx \left(\frac{LI^2}{2} + \frac{CV^2}{2} \right) = \frac{m\dot{q}^2}{2} + \frac{m\omega_r^2 q^2}{2}. \tag{5.26}$$

This fully coincides with the Hamiltonian of a harmonic oscillator. This allows the system to be quantized with the generalized coordinate q and the conjugate momentum $p = m\dot{q}$. It is convenient to introduce the annihilation and creation operators

$$a(t) = \frac{m\omega_r q + ip}{\sqrt{2m\hbar\omega_r}}, \quad a^\dagger(t) = \frac{m\omega_r q - ip}{\sqrt{2m\hbar\omega_r}}. \tag{5.27}$$

Comparing Eqs. (5.24) and (5.17), note that $a(t) = e^{i\omega t}a$ and $a^\dagger(t) = e^{-i\omega t}a^\dagger$. These operators act in the space of the number of photons — these are the so-called *Fock states*,

$$a|n\rangle = \sqrt{n}|n-1\rangle, \quad a^\dagger|n-1\rangle = \sqrt{n}|n\rangle. \tag{5.28}$$

In terms of these operators, we can rewrite the operators of the current and voltage and the Hamiltonian as follows

$$I = I_{r0}(a + a^\dagger)\cos\frac{\pi x}{l}, \quad I_{r0} = \sqrt{\frac{\hbar\omega_r}{L_r}},$$

$$V = iV_{r0}(a - a^\dagger)\sin\frac{\pi x}{l}, \quad V_{r0} = \sqrt{\frac{\hbar\omega_r}{C_r}}, \tag{5.29}$$

$$H_r = \hbar\omega_r\left(a^\dagger a + \frac{1}{2}\right).$$

In particular, we obtain that on the boundaries, $x = \pm l/2$, there is no current and the voltage equals to $\pm W$, with $W = iV_{r0}\langle a - a^\dagger\rangle = -2V_{r0}\,\mathrm{Im}\,\langle a\rangle$. So, we have related the voltage on the ends of the resonator to the mean value of the operator of the photon field in the resonator.

5.2.3. *Hamiltonian of the system "qubit-resonator"*

As we have seen, the qubit Hamiltonian in the flux basis (which is the basis of the circulating-current states) $\{|\uparrow\rangle, |\downarrow\rangle\}$ has the form $H_{qb} = -\frac{\Delta}{2}\sigma_x - \frac{\varepsilon_0}{2}\sigma_z$. The qubit current operator in this basis equals to $I_{qb} = -I_p\sigma_z$. The qubit and the resonator are connected inductively through the mutual inductance M, and hence the Hamiltonian of their interaction has the form:

$$H_{int} = MI(0)I_{qb} = -\mathrm{g}(a^\dagger + a)\sigma_z, \quad \mathrm{g} = MI_{r0}I_p. \tag{5.30}$$

The full Hamiltonian of the system (without taking into account the relaxation processes and excitation) includes the contributions of the non-interacting qubit, the resonator and the interaction term:

$$H_{\text{qb-r}} = H_{\text{qb}} + H_{\text{r}} + H_{\text{int}}$$
$$= -\frac{\Delta}{2}\sigma_x - \frac{\varepsilon_0}{2}\sigma_z + \hbar\omega_r\left(a^\dagger a + \frac{1}{2}\right) - \text{g}(a^\dagger + a)\sigma_z. \tag{5.31}$$

This Hamiltonian is analogous to the one, which describes the interaction of atoms and photons in *cavity quantum electrodynamics* [Scully and Zubairy 2012]. By analogy, the part of the solid-state theory where we deal with similar systems, is called *circuit quantum electrodynamics* (cQED).[39]

Let us change to the representation of the qubit eigen-states, similar to how we did this before. This is needed, for example, in consideration of the dissipative processes; in particular, the qubit relaxation appears from the excited to the ground state. Namely, let us make the transformation

$$S = \exp\left(i\frac{\eta}{2}\sigma_y\right), \quad \sin\eta = \frac{\Delta}{\Delta E},$$
$$\cos\eta = -\frac{\varepsilon_0}{\Delta E}, \quad \Delta E = \sqrt{\Delta^2 + \varepsilon_0^2}, \tag{5.32}$$

such that $S^\dagger H_{\text{qb}}S = \frac{\Delta E}{2}\sigma_z$. Then in the new representation, the full Hamiltonian takes the form

$$H'_{\text{qb-r}} = \frac{\Delta E}{2}\sigma_z + \hbar\omega_r\left(a^\dagger a + \frac{1}{2}\right) - \text{g}(a^\dagger + a)\left(\frac{\varepsilon_0}{\Delta E}\sigma_z - \frac{\Delta}{\Delta E}\sigma_x\right). \tag{5.33}$$

We now introduce the operators $\sigma_\pm = \frac{1}{2}(\sigma_x \pm i\sigma_y)$, which can be interpreted as the raising/lowering operators for the qubit:

$$\sigma \equiv \sigma_- = \frac{1}{2}(\sigma_x - i\sigma_y) = \begin{pmatrix} 0 & 0 \\ 1 & 0 \end{pmatrix},$$

---

[39]R. J. Schoelkopf and S. M. Girvin, Wiring up quantum systems, Nature **451**, 664 (2008).

$$\sigma^\dagger \equiv \sigma_+ = \frac{1}{2}(\sigma_x + i\sigma_y) = \begin{pmatrix} 0 & 1 \\ 0 & 0 \end{pmatrix}, \tag{5.34}$$

$$\sigma\,|e\rangle \equiv \sigma\,|1\rangle = \sigma\begin{pmatrix} 1 \\ 0 \end{pmatrix} = \begin{pmatrix} 0 \\ 1 \end{pmatrix} = |0\rangle \equiv |g\rangle, \quad \sigma^\dagger\,|0\rangle = |1\rangle,$$

or, in unified form:

$$\sigma|k\rangle = \sqrt{k}|k-1\rangle, \quad \sigma^\dagger|k-1\rangle = \sqrt{k}|k\rangle, \quad k = 0, 1 \tag{5.35}$$

which is analogous to Eq. (5.28). Note that the vector state of the system can now be written in the form:

$$|k, n\rangle = |k\rangle \otimes |n\rangle, \quad k = \{0, 1\}, \quad n = \{0, 1, 2, \ldots\}. \tag{5.36}$$

Let us now put $\sigma_x = \sigma + \sigma^\dagger$ in Eq. (5.33). We leave only the terms, corresponding to saving the energy in the system. Then we obtain the famous *Jaynes–Cummings Hamiltonian*:

$$H_{\mathrm{JC}} = \frac{\Delta E}{2}\sigma_z + \hbar\omega_r\left(a^\dagger a + \frac{1}{2}\right) + g_\varepsilon(a^\dagger\sigma + a\sigma^\dagger), \quad g_\varepsilon = g\frac{\Delta}{\Delta E}. \tag{5.37}$$

This procedure means that we have neglected the terms proportional to a, $a^\dagger\sigma^\dagger$ and so forth. Note that this approximation is equivalent to the rotating-wave approximation, which assumes $\Delta E/\hbar \approx \omega_r$. In order to confirm this, the interested reader can change to the interaction representation as follows

$$|\psi'(t)\rangle = \exp\left(\frac{i}{\hbar}H_0 t\right)|\psi\rangle, \quad H_0 = \frac{\Delta E}{2}\sigma_z + \hbar\omega_r a^\dagger a. \tag{5.38}$$

The interaction term in Eq. (5.37) describes the transfer of a photon to the resonator ($a\sigma^\dagger$) and the other way round ($a^\dagger\sigma$).

Let us describe now the excitation of such a system. For this, we assume that the alternating signal is applied at the resonator input, which is to the left capacitance in Fig. 5.2,

$$V_{\text{in}} = V\big|_{x=-\frac{l}{2}-0} = V_A \sin \omega_d t = \frac{V_A}{2i}\left(e^{i\omega_d t} - e^{-i\omega_d t}\right). \tag{5.39}$$

To the right, the voltage is given by the operator in Eq. (5.29):

$$V_{\text{res}} = V\big|_{x=-\frac{l}{2}+0} = iV_{r0}\left(a - a^\dagger\right)\sin\frac{\pi x}{l}\bigg|_{x=-l/2} = -iV_{r0}\left(a - a^\dagger\right). \tag{5.40}$$

The energy, corresponding to such microwave excitation is

$$H_{\mu w} = \frac{C_0 \Delta V^2}{2} = \frac{C_0}{2}\left(V_{\text{in}}^2 + V_{\text{res}}^2 - 2V_{\text{in}}V_{\text{res}}\right). \tag{5.41}$$

After omitting the "fast-rotating" terms and constants in this Hamiltonian, in the rotating-wave approximation, we obtain

$$H_{\mu w} = \xi\left(a^\dagger e^{-i\omega_d t} + a e^{i\omega_d t}\right), \quad \xi = \frac{1}{2}C_0 V_A V_{r0}. \tag{5.42}$$

This Hamiltonian has to be added up with the one in Eq. (5.37), so as to obtain the total Hamiltonian for the system "qubit-resonator" while accounting for the excitation.

5.2.4. *Coherent states and the quasi-classical Hamiltonian*

Let us now describe the photons in the resonator by means of the so-called *coherent states*. We will first learn this notion briefly, referring to [Scully and Zubairy 2012]. A coherent wave packet has minimal uncertainty and is analogous to a classical field. The coherent state $|\alpha\rangle$ is defined as the eigen-state of the annihilation operator a and can be obtained from the vacuum state by the action of the operator D as follows

$$a|\alpha\rangle = \alpha|\alpha\rangle, \quad |\alpha\rangle = D|0\rangle \equiv \exp(\alpha a^\dagger + \alpha^* a)|0\rangle. \tag{5.43}$$

The operator D is the unitary one, $D^\dagger = D^{-1}$, and acts as the displacement operator,

$$D^{-1}aD = a + \alpha, \quad D^{-1}a^\dagger D = a^\dagger + \alpha^*. \tag{5.44}$$

The value α describes the mean number of photons in the coherent state:

$$\langle n \rangle = \langle \alpha | a^\dagger a | \alpha \rangle = \langle 0 | D^{-1} a^\dagger D D^{-1} a D | 0 \rangle = \langle 0 | (a^\dagger + \alpha^*)(a + \alpha) | 0 \rangle$$
$$= |\alpha|^2. \tag{5.45}$$

Note that the coherent states can be expressed via the Fock states:

$$|\alpha\rangle = e^{-|\alpha|/2} \sum_{n=0}^{\infty} \frac{\alpha^n}{\sqrt{n!}} |n\rangle, \tag{5.46}$$

and the probability that there are n photons in the state $|\alpha\rangle$ is described by the Poisson distribution:

$$p(n) = \frac{\langle n \rangle^n e^{-\langle n \rangle}}{n!}. \tag{5.47}$$

Before averaging over the coherent state, let us make the transformation to get rid of the corresponding temporal term: $U = \exp\left(i\omega_d t a^\dagger a\right)$. We note that $i\hbar \dot{U} U^\dagger = -\hbar\omega_d a^\dagger a$ and $U a U^\dagger = a e^{-i\omega_d t}$. Then the total Hamiltonian, before and after the transformation (see Eqs. (5.31) and (5.42)), can be written as follows

$$H_{\text{tot}} = -\frac{\Delta}{2}\sigma_x - \frac{\varepsilon_0}{2}\sigma_z + \hbar\omega_r a^\dagger a - g(a^\dagger + a)\sigma_z$$
$$+ \xi\left(a^\dagger e^{-i\omega_d t} + a e^{i\omega_d t}\right),$$

$$\tilde{H} = U H_{\text{tot}} U^\dagger + i\hbar \dot{U} U^\dagger \tag{5.48}$$

$$= -\frac{\Delta}{2}\sigma_x - \frac{\varepsilon_0}{2}\sigma_z + \hbar(\omega_r - \omega_d) a^\dagger a - g(a^\dagger e^{i\omega_d t} + a e^{-i\omega_d t})\sigma_z$$

$$+ \xi\left(a^\dagger + a\right).$$

We average now the Hamiltonian over the coherent state, and we obtain the Hamiltonian of the qubit interacting with the resonator in the coherent state:

$$H = \langle \tilde{H} \rangle = \langle \alpha | \tilde{H} | \alpha \rangle = -\frac{\Delta}{2}\sigma_x - \frac{\varepsilon_0}{2}\sigma_z + \hbar(\omega_{\rm r} - \omega_{\rm d})\langle n \rangle$$
$$- {\rm g}(\alpha^* e^{i\omega_{\rm d}t} + \alpha e^{-i\omega_{\rm d}t})\sigma_z + \xi(\alpha^* + \alpha). \tag{5.49}$$

Now, let α be a real value (its phase can be excluded by shifting the initial moment of time, see the term with g). Then, omitting the constants, we finally obtain the Hamiltonian for the qubit connected to the driven resonator:

$$H = -\frac{\Delta}{2}\sigma_x - \frac{\varepsilon_0 + A_{\rm d}\cos\omega_{\rm d}t}{2}\sigma_z, \quad A_{\rm d} = 4\alpha{\rm g} = 4\sqrt{\langle n \rangle}{\rm g}. \tag{5.50}$$

So, we note, with pleasure and satisfaction, that we came to the Hamiltonian of a driven qubit, which we actually considered in the previous chapters.

5.3. Hybrid systems on the base of mesoscopic ones

The previous two sections were devoted to the mesoscopic systems on the base of superconducting Josephson circuits and the normal low-dimensional conductors. These systems have a number of advantages. For example, the opportunity of controlling the parameters in the wide range. To the disadvantages of these systems, we face the challenges of isolating such quantum systems from the environment. Absent of these disadvantages are the *micro*scopic systems, to which we can relate electrons, photons, atoms. What is interesting is the opportunity of linking the microscopic and mesoscopic subsystems (see Footnote 9 on page 15). For example, for applications in the field of quantum computations, mesoscopic qubits can be used in the quantum processor, and microscopic qubits can be used in the capacity of the long-term memory.

For mesoscopic quantum systems, in distinction from their microscopic counterparts, it is important to take into account the impact of controlling and read-out devices. The control can be executed by means of applying current, voltage, or magnetic flux. These

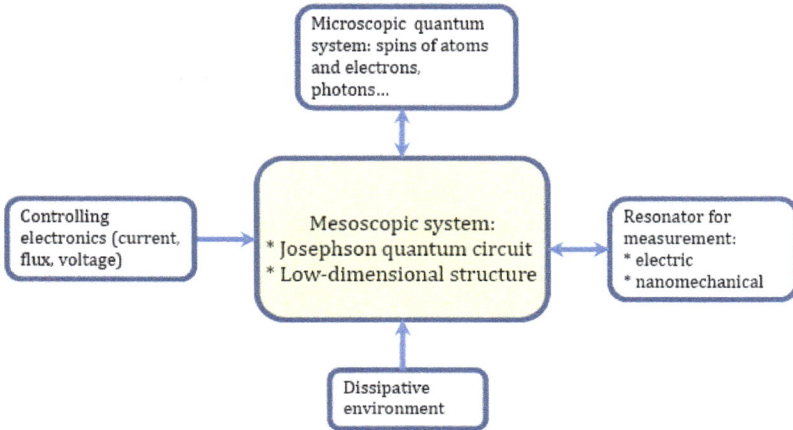

Fig. 5.3. Schematic of a hybrid system on the base of a mesoscopic subsystem.

parameters can have both constant and alternating components. For reading out the states, one can use the resonators, in particular, nanomechanical or electrical ones, as considered in the previous two sections.

The relatively large size and additional electronics result in the essential influence of the dissipative environment on the mesoscopic system. This issue was discussed when we considered dissipative dynamics in Chapter 2.

Finally, a mesoscopic system itself may not consist singly of one but a system of qubits, or from a qubit coupled to a quantum resonator. Partly, we have addressed such problems in the present Chapter. In summary, we can unify what was said in the form of a schematic; see Fig. 5.3.

Conclusion to Chapter 5

Quantum optics or quantum electrodynamics of optical resonators studies the fundamental interaction of atoms and electromagnetic field. The field is characterized by the frequency ω_0 and the energy of photons $\hbar\omega_0$. There are two possible situations: when $\hbar\omega_0 \ll k_B T$, then the field can be considered in the semi-classical approximation, and when $\hbar\omega_0 \geq k_B T$, then the entire system should be considered as an aggregate quantum-mechanical object.

In this Chapter, we have considered the solid-state realizations of such systems, where the role of an atom is played by a normal or superconducting mesoscopic subsystem, the qubit, and the oscillator can be a nanomechanical or electrical resonator. On these specific examples, we have demonstrated how to describe such systems in the two cases. In particular, a qubit-resonator system is described by the Jaynes-Cummings Hamiltonian; in the semi-classical approximation this reduces to the Hamiltonian (5.50) for a driven qubit. So, we considered the basic approaches and some of the notions of this new field — circuit quantum electrodynamics.

Problems for independent work and for self-assessment

5.1. (*****) With reference to Footnote 36 on page 138, study how oscillations in a non-linear system can be reduced to the oscillations of an equivalent linear system. This is a very useful task and pedagogical trick for a theoretical physics student, though this can be skipped for the first reading of the lecture course.

5.2. (***) Lead the electrostatic analysis of a quantum dot, formed by the capacitors, of which one is variable (e.g. due to a mechanical resonator), and demonstrate that this can be described by the parametric capacitance, Eq. (5.5).

5.3. (***) Expanding in series, for small displacement u, demonstrate that the resonator frequency shift is defined by the quantum subsystem state, as in Eq. (5.10).

5.4. (**) Find the quantum capacitance $C_Q \propto \partial \langle n \rangle / \partial n_g$ for the charge qubit.

5.5. (**) Describe a transmission line; obtain the telegraph equation with losses.

5.6. (***) Obtain and quantize the Hamiltonian of a transmission-line one-mode resonator.

5.7. (*) Write down the Hamiltonian of a coupled qubit and a quantum oscillator, Eq. (5.31).

5.8. (**) Assuming the rotating-wave approximation, transform the Hamiltonian from the previous task to the Jaynes–Cummings Hamiltonian.

5.9. (*) Include driving via voltage to the qubit-resonator Hamiltonian.

5.10. (***) Average the qubit-resonator Hamiltonian over a coherent state, and demonstrate that the Jaynes–Cummings Hamiltonian is then reduced to the semi-classical Hamiltonian of a driven qubit, Eq. (5.50), which we used in previous chapters.

CONCLUSION

"However, all of what I depicted..., apparently
with such needless particulars — all this leads
to the future and will be necessary there. At
its place, everything will respond; I could not
refrain; and if boring, I would ask not to read."

F. M. Dostoevskiy[40]

In conclusion of this lecture course, let us summarize principal
definitions and results, which were considered in the respective
chapters.

(0) Mesoscopic Physics (Mesoscopics) studies the manifestations of
quantum effects in systems of many particles (condensed media)
on scales and time spans when the system phase coherence is
important. Having in view possible applications, this field of
physics is also called Quantum Engineering.

(1) Quantum computation is the field of studying problems related
to the manipulation and transmission of information using the
laws and objects of quantum mechanics. The theory of quantum
computation can be seen as a subdivision of non-relativistic

---

[40]F. M. Dostoevskiy, The Adolescent, Literatura artistica, Chisinau (1986);
the quotation was translated here by the author {this quotation in Russian:
«Впрочем, и все, что описывал..., по-видимому с такой ненужной
подробностью, – все это ведет в будущее и там понадобится. В своем
месте все отзовется; избежать не умел; а если скучно, то прошу не
читать.»}.

quantum mechanics, of which the central object is a two-level system — the qubit, for brevity. A qubit is described by a wave function or a state vector, which corresponds to the superposition of two basis states:

$$|\psi\rangle = \alpha|0\rangle + \beta|1\rangle. \tag{i}$$

The principle of superposition, generalized for a system of two qubits, results in the entangled states, when the wave function, e.g. $(|01\rangle - |10\rangle)/\sqrt{2}$, cannot be factorized. This corresponds to the non-local quantum correlation, the manifestation of which is sometimes referred to as the Einstein-Podolsky-Rosen paradox. These correlations present the basic resource for quantum computation. And their fundamental importance is emphasized by the fact that quantum correlations are stronger than classical ones ($S_q > S_{cl}$), which is known as the violation of the Bell inequalities.

(2) Many important problems can be considered by studying the dynamics of a two-level system with periodic excitation, i.e. the driven qubit, which is described by the Hamiltonian

$$H(t) = -\frac{\Delta}{2}\sigma_x - \frac{\varepsilon(t)}{2}\sigma_z, \quad \varepsilon(t) = \varepsilon_0 + A\cos\omega t. \tag{ii}$$

The distance between the stationary energy levels of the qubit, $\Delta E = \sqrt{\Delta^2 + \varepsilon_0^2}$, is essentially the controllable value. When the excitation frequency ω is a multiple of the qubit characteristic frequency $\omega_{\text{qb}} = \Delta E/\hbar$, namely when $k \cdot \omega = \omega_{\text{qb}}$, then we have the multi-photon excitation of a qubit, related to the absorption of k photons of the driving field by the two-level system. In particular, when the frequencies are equal, $\omega = \omega_{\text{qb}}$, the occupation of the qubit levels oscillates between 0 and 1 with the Rabi frequency Ω_R ($\Omega_R \propto A$): $P_+(t) = (1/2)(1 - \cos\Omega_R t)$. In the inverse limiting case of adiabatically slow changes, when $\omega \ll \omega_{\text{qb}}$ — the transitions between the levels are described by the Landau–Zener–Stueckelberg–Majorana formula: $P_+ = \exp\left(-\frac{\pi\Delta^2}{2A\hbar\omega}\right)$. The phase accumulation under such evolution, $\zeta = \int dt\sqrt{\Delta^2 + \varepsilon(t)^2}/2\hbar$, results in interference phenomena — the Stückelberg oscillations.

(3) Many applications of superconductivity are related to the Josephson effects. The stationary Josephson effect consists of a non-dissipative supercurrent flowing through the tunnel junction, $I_J = I_c \sin \phi$. Here, the phase difference ϕ in a doubly-connected geometry is defined by the external magnetic flux: $\phi = 2\pi\Phi_e/\Phi_0$, where $\Phi_0 = hc/2|e|$ is the flux quantum. And if the direct voltage is applied to the contact, then the phase difference depends on time, $\dot{\phi} = 2eV/\hbar$, and the current oscillations appear with the frequency $\omega = 2|eV|/\hbar$ — this is the non-stationary Josephson effect. A Josephson junction is characterized by a non-linear inductance, $L_J = L_J(\phi)$, and this allows qubits to be realized on this, which makes up the procedure of quantizing respective circuits. For the phase, charge, and flux qubits, which are described by the Hamiltonian (ii), their parameters depend on the applied current, voltage, and magnetic flux, respectively:

$$\text{phase: } \varepsilon = \varepsilon(I_{\text{bias}}), \quad \text{charge: } \varepsilon = \varepsilon(V_{\text{g}}), \quad \text{flux: } \varepsilon = \varepsilon(\Phi_{\text{e}}). \quad \text{(iii)}$$

And so, the superconducting circuits result in the appearance, on the mesoscopic scale, of not only coherent effects such as the flowing of non-dissipative currents, but also in the appearance of the superposition states.

(4) If the size of a conducting system is comparable with the Fermi wavelength, $L_i \sim \lambda_F$, then this becomes a low-dimensional system — a two-dimensional electron gas, a quantum wire, or a quantum dot. In this case, the quantization of charge and spectrum, as well as the interference, becomes important. The conductivity of a one-dimensional conductor — the wire — is proportional to the transmission coefficient: in particular, at zero temperature

$$G|_{T=0} = G_0\tilde{T}(\mu). \quad \text{(iv)}$$

And even at the reflectionless transmission, at $\tilde{T} = 1$, the conductance is finite and equal to the conductance quantum $G_0 = 2e^2/h$. For mesoscopic conductors, the Aharonov–Bohm effect is important: the vector and scalar potentials, $\mathbf{A}$ and φ, directly influence the observable values. In particular, the conductance of a doubly-connected sample depends on the piercing magnetic flux: $G(\Phi) \sim \cos\left(2\pi\Phi/\Phi_0^{(N)}\right)$. Here, the periodicity is described by the value $\Phi_0^{(N)} = hc/|e|$, which differs from the superconducting magnetic flux quantum by the factor of 2.

(5) Modern mesoscopic physics also studies composite systems. Such a hybrid system, on the base of a mesoscopic quantum subsystem, can also include a microscopic subsystem, controlling electronics, read-out resonators, and dissipative environment. We have considered two illustrative examples: a quantum dot plus a classical nanomechanical resonator and a flux qubit plus a quantum electrical resonator. In the former case, one can define the average number of electrons on the dot $\langle n \rangle$ from the resonator frequency shift $\Delta\omega$. If a resonator frequency is sufficiently large, $\hbar\omega_r \geq k_B T$, then it should be described quantum-mechanically. For an electric resonator, in this case, the current and the voltage are replaced by the photon creation and annihilation operators, and the resonator's Hamiltonian is $\hbar\omega_r \left(a^\dagger a + 1/2\right)$. Then the qubit-resonator system is described by the Jaynes–Cummings Hamiltonian

$$H = \frac{\Delta E}{2}\sigma_z + \hbar\omega_r a^\dagger a + \mathrm{g}\left(a^\dagger\sigma + a\sigma^\dagger\right). \qquad \text{(v)}$$

If a resonator is classical, it is described by means of the coherent states and then the Hamiltonian of the form (v) reduces to the Hamiltonian (ii). In the general case, the system qubit-resonator is formally analogous to the basic system of quantum optics — the atom-photon system, describing the interaction of matter and field. The branch of physics studying such systems is called circuit quantum electrodynamics.

Acknowledgments

I am grateful to I. V. Krive, M. V. Moskalets, F. Nori, A. M. Zagoskin for useful and stimulating comments. The many problems considered here have been put through long study together with my scientific advisors A. S. Rozhavsky, Yu. A. Kolesnichenko, and A. N.

Omelyanchouk. For my interest and understanding of physics of mesoscopic structures, I am grateful to the communication with the colleagues and co-authors: S. Ashhab, O. Astafiev, V. Cherkaskiy, A. Fedorov, M. F. Gonzalez-Zalba, M. Grajcar, Ya. S. Greenberg, E. Il'ichev, W. Krech, F. Nori, G. Oelsner, K. Ono, V. I. Shnyrkov, A. M. Zagoskin and others.

The writing of this lecture course was initiated by the kind proposition of the Head of the theoretical department in the Faculty for Physics and Technology of Kharkov National University, V. D. Khodusov. My interest in delivering lectures there, which resulted in this textbook, is to a large degree caused by an interest in the lectures from students of the groups TYa-51 during the years from 2014 to 2019.

Finally, I am grateful to various members of World Scientific Publishing Company, especially Soh Yong Qi, Rachel Seah Mei Hui and Christopher B. Davis, who have assisted in the publication of this book.

BIBLIOGRAPHY

(for further reading, besides the references appearing in the main text as footnotes)

[Blum 1981]
   K. Blum, Density Matrix Theory and Applications (Plenum, NY, 1981); (Mir, Moscow, 1983).
[Valiev 2005]
   K. A. Valiev, Quantum computers and quantum computations, Physics-Uspekhi **48**, 1 (2005); UFN **175**, 3 (2005).
[Greenstein and Zajonc 2006]
   G. Greenstein and A. G. Zajonc, The Quantum Challenge: Modern Research on the Foundations of Quantum Mechanics (Jones and Bartlett Publishers, Sudbury, MA, 2006); (Intellect, Moscow, 2008).
[Zagoskin 2011]
   A. M. Zagoskin, Quantum Engineering: Theory and Design of Quantum Coherent Structures (CUP, Cambridge, 2011).
[Landau and Lifshitz 1977]
   L.D. Landau & E.M. Lifshitz, Quantum Mechanics: Non-Relativistic Theory (Pergamon Press, Oxford, 1977); (Nauka, Moscow, 1989).
[Moskalets 2010]
   M. V. Moskalets, Introduction in Mesoscopic Physics (NTU KhPI, Kharkov, 2010).
[Nielsen and Chuang 2010]
   M. A. Nielsen and I. L. Chuang, Quantum Computation and Quantum Information (CUP, Cambridge, 2010); (Mir, Moscow, 2006).
[Penrose 2003]
   R. Penrose, The Emperor's New Mind: Concerning Computers, Minds, and the Laws of Physics (Oxford University Press, Oxford, 2002); (URSS, Moscow, 2003).

[Scully and Zubairy 2012]
    M. O. Scully and M. S. Zubairy, Quantum Optics (CUP, Cambridge, 2012); (Fizmatlit, Moscow, 2003).
[Schmidt 1997]
    V. V. Schmidt, The Physics of Superconductors (Springer-Verlag, Berlin Heidelberg, 1997); (MCNMO, Moscow, 2000).

INDEX

list of terminologies of which their study is one of the major aims of this lecture course